ISLE ROYALE N.P.

ACADIA N.P.

CAPE COD N.S.

MONOMOY N.W.R.

FIRE ISLAND N.S.

Mississippi River

Ohio River

BLUE RIDGE PARK

SHENANDOAH N.P.

MAMMOTH CAVE N.P.

GREAT SMOKY MOUNTAINS N.P.

CAPE HATTERAS N.S.

HOT SPRINGS N.P.

Atlantic Ocean

OKEFENOKEE N.W.R.

Atchafalaya Basin

MARSH ISLAND

WEEKI WACHEE SPRINGS

Lake Okeechobee

Corkscrew Swamp

Gulf of Mexico

EVERGLADES N.P.

Our
Vanishing
Wilderness

Our Vanishing Wilderness

by
Mary Louise and Shelly Grossman
and
John N. Hamlet

Photographs by Shelly Grossman

MADISON SQUARE PRESS

GROSSET & DUNLAP, PUBLISHERS, NEW YORK

Also by the authors:

BIRDS OF PREY OF THE WORLD

THE STRUGGLE FOR LIFE IN THE ANIMAL WORLD

Book design by SHELLY GROSSMAN

End paper map by BUNJI TAGAWA

Editor: NORTON WOOD

1972 Printing
Copyright © 1969 by Grosset & Dunlap, Inc.
All rights reserved
Library of Congress catalog card number: 68–26147
ISBN: 0–448–01208–1

Published simultaneously in Canada

Printed in the United States of America

Contents

A butterfly feeds on nectar . . .

A dragonfly seizes the butterfly . . .

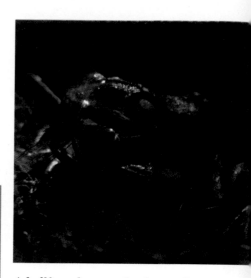

A bullfrog devours the dragonfly . . .

Introduction

A snake swallows the bullfrog . . .

And a red-shouldered hawk flies off with the snake.

The zebra swallowtail butterfly takes nectar from an evening primrose. The butterfly is caught by a dragonfly, which is captured by a bullfrog, which is attacked and swallowed by a water snake, which ends up as food for a red-shouldered hawk. When this predator dies, its body will decompose and its minerals and chemicals will return to the soil, allowing new plant growth to continue. The sequence is one small event in the vast cycle of life on earth. A continuous inflow of energy from the sun keeps the cycle going. From the moment oxygen first became available as a by-product of plant photosynthesis, at the beginning of the world as we know it, the entire evolution of living organisms can be seen as a series of elaborations of the food chain. Heat energy, manufactured from light, is distributed to plant-eaters, which are preyed upon by meat-eaters.

Each member of the chain uses most of the energy consumed just to live and reproduce; the leftover energy that gets passed on if an animal is eaten is relatively little. This is an example of what physicists know as the second law of thermodynamics: because of heat dispersion, no spontaneous transformation of energy from one form into another can ever be 100 per cent efficient.

Energy loss through long food chains is a kind of natural taxation, a payment for stability in natural communities. It is in the most complex of these communities, such as the primarily deciduous forests of the Temperate Zone, that food chains are longest; together they form a web of life that can repair itself when strands are broken. In a few thousand acres of forest, there may be a dozen or so different tree communities, some dominated by one kind of tree and others dominated by two or three different trees with their associated plants and animals. If disease destroys one species of tree, another moves in to replace it—a phenomenon demonstrated when a foreign fungus wiped out all the American chestnuts from the Appalachian ridges, and other hardwoods quickly grow up to fill the void. If one crop of nuts fails in a season, there are other foods for the nut-eaters to switch to, and their predators will not go hungry. Thus, with few exceptions, animal populations remain constant from year to year, and fluctuate relatively little over periods of many years.

When fire or wind destroys a forested area, there is a gradual development of new cover and new sources of food, from sunny meadow to scrub to shaded forest once again—the complete cycle taking from eighty to a hundred years. The cast of animal characters enlarges with each stage in the forest rebuilding, a different variety of life existing at each level from the ground up to the highest treetops. The multiplicity of living-places and of food available gives stability to the forest habitat.

By contrast, the cold tundra of the far North is a simple community of short food chains and instability. On the monotonous heath, underlain by permafrost and dotted with ponds and lakes, variety exists only among the water birds. These species breed during a few weeks in June and July and then migrate south. The short summer limits plant growth and development; the colonization of rock by low-lying lichens may take hundreds of years. Taller, shrubby species of lichens support herds of caribou, or reindeer, that must wander far to feed because the range recovers so slowly after grazing. The constant movement of the herds, the long migrations to avoid the cold, the wolves that follow, all help to prevent overpopulation, which could mean starvation.

The problem of overpopulation was dramatically illustrated by the introduction of 25 reindeer onto St. Paul Island, in the Pribilofs, in 1911. On this limited range, free of wolves, the reindeer at first prospered. The herd grew to 2,046 in 27 years; then, in just 12 years, it dropped to eight. From a study of the St. Paul Island reindeer, biologists determined that each animal needed a year-round grazing range of at least 33 acres—three times more than was available per deer on the island just before the population "crash."

The populations of small rodents that feed on the berries and seeds of the tundra are notoriously unstable. Every three or four years the lemmings multiply by as much as 500 per cent and then, inexplicably, crash. Arctic foxes, hawks and owls also increase in numbers—always after the peak of the prey cycle—and consequently there is for a time a massive scarcity of food. The arctic foxes starve. Snowy owls drift as far south as Bermuda in the Western Hemisphere, and to France, Austria and the Balkans in Europe. Nowhere is this cyclic imbalance more drastically evident than in Norway, where lemmings periodically crowd together in a condition of mass hysteria and migrate, one by one, to their death in the sea. But similar explosions occur in cycles at different times in various parts of Hudson Bay and in the far Northwest of the American tundra.

Over the million-year period of the Pleistocene Epoch, nearly all of the land that is now forested in the eastern United States and in Central Europe was repeatedly covered by glaciers, and each retreat of the ice left only tundra. Today many arctic plants grow on the cold peaks of high mountains in the temperate zone, a relict flora that is hauntingly similar in the Alps, the Pyrenees, the Rocky Mountains and other chains all across the Northern Hemisphere. Not only are alpine flowers present in the high regions of North America, but also animal migrants from Eurasia—the marmot, the

Predator in the Spruce
A lynx ranges through an Eastern coniferous forest on big, silent paws. The numbers of this carnivore vary with the populations of rodents and other prey.

pika, the mountain sheep (mouflon) and the mountain goat (closely related to the chamoix), all of which crossed the Bering Land Bridge by the end of the Ice Ages, enriching the fauna of this most glaciated of the continents.

Between the extremes of deciduous forest and tundra, the North Temperate Zone of the earth contains numerous other environments of varying complexity and stability—coniferous forests, coastal pine woods and marshes, sandy and rocky shores, grasslands, dry cold steppes and hot deserts. The climates of all these places are determined not only by latitude but by prevailing winds and ocean currents, rainfall, and often by the mountain barriers that prevent rain from reaching the interiors of continents.

Mountains are a symptom of our time, as time is measured by the geologist from the late Cenozoic Era to the present. We are in the midst of one of the great periods of crustal unrest. Such periods are believed to have occurred only twice before in the five- or six-billion-year history of the earth, during which the "norm" seems to have been mild, tropical climates and vast seas. Now continents are larger than in previous eras, mountains are more numerous and stand higher, there are more volcanoes, and earthquakes are more frequent; Greenland and the Antarctic lie deeply covered with ice, and the valley glaciers of the Temperate Zone mountains spasmodically shrink and surge forward.

Much of this activity apparently comes as a result of gigantic rifts in the earth's crust, extending from the ocean floors far beneath the continents in certain places. These great fissures account for the Rift Valleys of East Africa, the Rhone River Valley, and the unique basin and range topography of the interior sagebrush country and hot deserts of western North America. Rifts also account for a string of volcanoes ringing the Pacific Ocean, from Japan and the South Seas islands to Peru, Mexico, California and the Valley of Ten Smokes in Alaska, as well as the volcanoes of East Africa, Italy and the Caribbean. Numerous islands have been formed by underseas

A Glacial Horn
A peak that forms part of the Continental Divide in northern Montana rises beyond the wide, flower-strewn valley cut by a huge glacier some 10,000 years ago.

eruptions along the mid-Pacific and mid-Atlantic Ridges—most recently the isle of Surtsey, off Iceland. Nowhere have the earth's inner fires broken through the planetary crust to paint a more dramatic picture than on the volcanic plateaus of the Yellowstone and among the Cascade Mountains of the American Northwest.

The relative position of mountain chains—running from north to south or from east to west—vitally affects the interiors of continents. All of western and southern Europe, for example, enjoys a mild maritime climate, both summer and winter. But a great portion of western and central North America is blocked off by high mountains from the effect of sea winds. To the eastward of Southern California's Coast Range, covered with "Mediterranean" forests and chaparral, is the Mojave Desert, the hottest and bleakest region of the continent.

Northern California's redwoods and the Sitka spruce of Oregon, Washington and British Columbia grow in a narrow strip along the coast, deriving most of their needed moisture from ocean fog—like the coastal forests on the Japanese island of Hokkaido. Almost tropical in their lush variety, and in the height to which they grow (the redwoods are the world's tallest trees), these forests provide a fantastic contrast to the cold sagebrush country on the leeward side of the Sierras and Cascades, in Nevada, Oregon and Washington.

The extremes of climate in the southwestern United States make it possible to see a "textbook" progression of vertical mountain zones within a relatively small geographical area, from hot desert at the base to cold, coniferous forest at the summit —something that is unique to this hemisphere. Farther north, in the Rocky Mountain area, the zones rise from sagebrush or grassland to Alpine tundra in a pattern more similar to the life zones of a European mountain. Most of Glacier National Park in the Montana Rockies lies above tree line, and its glacial "horns" and "cirques" are reminiscent of the Alps (though the big, Alpine-type glaciers of the American continent are farther north, in British Columbia and Alaska).

Through the millennia, the non-living environment has shaped the living organism to its outlines. Each habitat has developed its own food chains and food webs. These relationships are at the core of the science of ecology, which attempts to explain our total natural world: why plants and animals live in certain places, and only in certain combinations; why it is necessary to have more plant-eaters than carnivores; how plants and animals limit their own numbers in communities which are ever changing, from season to season and from year to year.

Only recently have we realized that a knowledge of ecology is invaluable to the survival of man in a largely man-made world. In the web of life, man is a newcomer. He arrived on the scene with the omnivorous habits of a raccoon, the ambition of a beaver to move the earth and stop the waters, and the unique cleverness of the human primate to invent tools for building and weapons with which to kill game and destroy any competitive meat-eaters. With the growth of civilization, man has abandoned his early hunting cultures and turned to farming. And in doing so he has destroyed vast natural areas.

At first the ranges needed for domestic cattle and the land for crops were a relatively insignificant proportion of the natural scene. There was room for the predators of the insects that ate the crops, for game, and for many other animals and birds. (To this day, the English hedgerow, the Swedish meadow-forest, the Western European coppice, and the southern Michigan woodlot and open field remain as examples of coexistence and relative harmony between man and nature.) When mechanized farming began, however, garden patches grew into vast monocultures from forty to a thousand or more acres in size. They brought an immediate increase in the population of rabbits and rodents, blackbirds, starlings, and most of all, insects. The pests were few in species, as on the tundra, but astronomical in numbers. Thus the new man-made habitats have evolved in the direction of great instability. They require constant manipulation. For the first time in history, it has been necessary to cope with lemming-like population explosions in the Temperate Zone— eruptions of the common vole in France, southern Germany and Russia, the field vole in England, the mountain and meadow voles in the United States, and the Levant vole in the Near East.

To Europeans and Americans steeped in the tradition of game protection, it has come as a shock in recent years to realize that over-protection of wildlife is destructive of both the wildlife and the land. The phenomenon may be seen in north Germany, in eastern Pennsylvania, on the Kaibab Plateau of Arizona, and in dozens of other, less publicized places. In each of these places an overabundance of deer, deprived of their natural enemies, made it impossible for certain of their food plants to survive and reproduce. Among the casualties have been beech, maple and yew in Europe, ground hemlock and white cedar in the eastern United States, and mountain mahogany and cliff-rose in the western United States. The result in each case has been a population crash

similar to that of the reindeer artificially introduced on St. Paul Island. If the browse and the deer recover, the cycle repeats itself again and again.

Although it is to our credit that the hottest of deserts can be made to bloom with man-made oases, it is also a matter of historical record that dunes and badlands have spread throughout the Mediterranean countries of Greece, Italy and Spain, and across parts of India and China, despite the prevalence in these places of forest climates. The new deserts are the work of man; they did not exist until hundreds of years of forest-cutting, livestock grazing and erosion had stripped

The Victim of Prejudice
Shot and hung on a Texas fence, a dead Swainson's hawk symbolizes man's misjudgment of nature. By preying on rabbits that destroy range grass, this bird actually benefits ranchers.

the countryside of its fertile topsoil.

From 1500 to the present, the semi-desert ranges of the American Southwest have been abused by Spanish, Mexican, and U.S. ranchers, with the result that mesquite has invaded the grama grasslands and all of the tobosa flats are now usurped by creosote bush. This tiny-leaved plant is of no use as browse. However, the wind-blown mound at its base houses the burrows of kangaroo rats and visiting side-winder rattle-snakes. (An equivalent to the creosote bush in the Arabian Desert helps to house the gerbil, as well as a side-winding viper.)

Northward, in the Great Basin (which is similar in many respects to the cold steppes of Siberia and the Gobi), overgrazing has destroyed the grasslands and promoted the spread of sagebrush. The common debate over whether grass ever existed in areas such as this can easily be resolved by visiting a wildlife refuge where grazing is allowed—but limited. The grass will usually be plentiful. Cross the border into Red Rocks Refuge in southern Montana, and suddenly the sagebrush is all brown. It has been sprayed with herbicides (2,4-D and 2,4,5-T) and presumably killed. As a result, the grass here has been increasing, though the roots of the sagebrush may not be dead.

It is true that the lost, and almost lost, grasslands are extremely sensitive to changes in climate. And the climate in these areas has become

The Painted Desert

Under a threatening sky, crumbling buttes and mesas in northeastern Arizona glow with color. Here summer floods are steadily washing the plains away to the Pacific.

hotter and drier in the last hundred years. But the change in climate has not caused—it has only contributed to—the advance of the desert occasioned by the activities of man.

Many of the so-called deserts of the United States are un-desertlike by Sahara standards. The typical American desert consists of an intermixing of low, hot basins with grassy plateaus and forested mountains. When seen by the first explorers and trappers, the river valleys here were marshy. The deep arroyos (gullies) were cut and the water table dropped at the time of a sharp rise in cattle grazing, toward the end of the 19th Century. Sheep and goats followed the cattle in a pattern all too familiar to observers of European ecological development. In the last thirty years,

the number of sheep and goats in West Texas alone has risen from a few thousand to six million head.

The changes often bring severe hardship to the original inhabitants of the land. Rodents and rabbits are poisoned by ranchers because they eat cattle forage, and predators are trapped and shot indiscriminately because they kill livestock. Too often a strictly local problem becomes an excuse to exterminate species on a broad scale. So it happened that 20,000 golden eagles were shot from airplanes in the southwestern United States in twenty years, whether or not they were killing sheep, and 77,258 coyotes were "controlled" by government trappers during one year alone (1966), sometimes in places where no sheep are

run. The ancient prejudice of man against the predator, which is world-wide—as strong in the herdsman on the Russian steppe as in the cowboy on the plains of Texas—ignores the root cause of the trouble, misuse of the land itself.

Ignorance can be destructive, and technology today is on a collision course with nature. One of the great tools of modern man in dealing with his environment is the insecticide. In twenty years, however, no less than 120 insect species have developed a resistance to DDT, and they keep outliving stronger and stronger chemical compounds. The battle to control insects on croplands and in commercial forests seems in danger of being lost unless we turn to a combination of natural with artificial means—the insecticides plus whatever living enemies of the insects can be encouraged or imported.

A potent reason for limiting the use of insecticides is the fact that they are not selective; besides insect pests, they kill other wildlife—birds, mammals, fishes—either by direct application, through aerial drift in chemical sprays, or through the runoff of residual poisons into the watercourses. Even "safe" levels of usage may not be nearly as safe as supposed. Chlorinated hydrocarbon compounds such as DDT, dieldrin and endrin break down slowly in animal bodies, and are passed along natural food chains in larger and larger concentrations that may eventually be lethal. Earthworms, for example, have been found to contain ten times more of an insecticide than the surrounding soil, and a hundred such worms are enough to kill a robin. A similar concentration process affects fertility; certain insecticides can be passed from parent to egg, resulting in a sterile egg or a dead chick. Insecticides can even bring about behavior. that is biologically destructive. The poisons in question affect the nervous system, and have been known to cause gulls and peregrine falcons to smash their own eggs.

DDT and similar compounds may be responsible for the nesting failures of bald eagles on Lake Michigan, in Maine, and in the mid-Atlantic

A Wild Coast Preserved
An early morning sun burns away mist enshrouding the coastline of Mt. Desert Island, Maine. This rugged shore is part of Acadia National Park, established in 1909.

states, and of the Osprey north of Chesapeake Bay. The peregrine falcon has virtually disappeared from the northeastern United States since the widespread introduction of insecticides, and has dropped to very low numbers on the other side of the Atlantic. In 1964 there were only about four pairs on the south coast of England and perhaps less than two hundred in all of Great Britain. No one knows why the American birds are gone, but the English birds have shown all the signs of nervous collapse that are now recognized as part of the DDT syndrome.

Although large mammals seem to be more capable of surviving this kind of environmental poisoning than a falcon, the loss of even a single species should be regarded as a warning signal. With pollution in all of its ugly forms spreading through major waterways, through the soil and even through the atmosphere of the earth, it is becoming more and more apparent that the victim of all this may ultimately be man himself.

By the year 2000 the human population of the earth will probably reach six billion. From a practical point of view, we stand now at a crossroads: the choice would appear to be coexistence with nature, or life in a synthetic world—if the creation of such a thing is possible. The most far-seeing of our technologists envision algae steaks for dinner and deep-sea housing developments. Because of the second law of thermodynamics (energy loss through long food chains), it surely would be more efficient to cultivate one-celled organisms in the sea than to raise cattle, which are multicellular organisms that must eat grass on land. Significantly, algae are also more resistant to atomic radiation than higher forms of plant life. Are we anticipating the bomb? Are we advancing or are we retreating into the sea?

Until the last decade, few Americans worried about the loss of wilderness or of its indigenous wildlife. From Daniel Boone to Lewis and Clark to the California '49ers and Buffalo Bill, the pioneers shot their way across the continent. The railroad builders lived off the great bison herds of the plains, and by 1900 the Army and the buffalo hunters had killed off the rest, except for a last few individuals in Yellowstone National Park. The present herds are descendants of these 21 survivors and a small herd in Canada—all that remained in place of an estimated population of 60 million animals in 1600. In the eastern United States, many species are regionally extinct or nearly so—wolves and cougars, wapiti (similar to the European red deer) and moose (the same animal as the European elk), the wood bison (closely related to the European wisent) and the holarctic golden eagle. In the last hundred years North America has lost 20 species, more than Europe has lost in the past 1,000 years. Fifty more species are on the U.S. Endangered List. This list, which was compiled recently by the federal government, includes California condors, whooping cranes, and grizzly bears—all creatures so rare that they may soon be gone forever. They are literally wards of the Bureau of Endangered Species, an agency established by Congress in 1966 to create special refuges and try to keep animals like this from dying out. Further evidence of the national concern for "rare" species is seen in the recently enacted Wilderness Bill, the proposed Wild Rivers Bill, and the large number of new National Parks created in both East and West.

A reason for all this public action is the phenomenal rise in the numbers of wildlife watchers the world over. In this country there are believed to be 25 or 30 million of them. Today the total of "outdoor" vacationers in the U.S. who do not fish or hunt nearly equals the total of those who do. A vast group of Americans belong to private conservation clubs; these are people who take vicarious pleasure in knowing that at least 51 California condors still exist, even though they will probably never see one. On the rise, too, is an enjoyment of nature as a whole, and an understanding of its management. Once an unknown word, "ecology" has acquired great importance in government planning at every level. Although much progress

lies necessarily in the future, it is clear now that a new order is on the way.

National Parks will no longer be kept like historic buildings through which the admiring public wanders—and wonders only at the tallest trees or the highest waterfalls. Recreational use will increase in National Forests that have long been managed (or mismanaged) primarily for one or two species of game or fish, or to satisfy lumbering, mining and grazing interests. The concept of a refuge will broaden; even today, the manager of a moose refuge finds himself increasingly involved with aspens and beavers, frogs and mice. Only the primitive and wilderness areas, reserved for the lonely pursuits of canoeing, packing, and hiking, will not be expanded in acreage. These pristine places—the roadless back country of America—can only shrink as the natural world is encroached upon, for whatever reason the encroachment.

Once changed, the wilderness cannot be legislated back into existence. It is because of this that the truly wild places are of increasing value to science. Often they contain combinations of plant and animal species that have been destroyed everywhere else. And each species in itself represents a unique combination of genes, never to be repeated when once extinct.

In this book—which represents 67,000 miles of travel, as well as four years of research and photography—the authors provide a look at some of the wonderfully varied natural communities of the continental United States, as a basis for understanding the ecology of the world we live in. Here is the story of how communities of plants and animals have changed over the ages, and of how they are being changed today—both by climatic influences and by the rising populations and technological assaults of man.

1
The Seasons

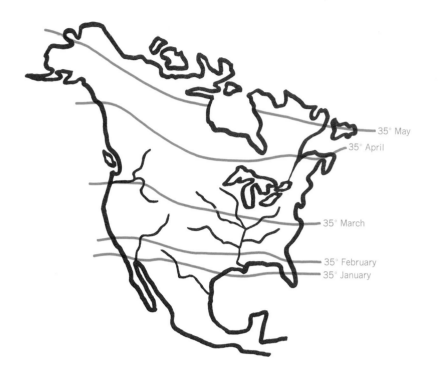

35° May
35° April
35° March
35° February
35° January

DANVILLE—*June 15, 1816*. It was late for such a fall of snow in Vermont. In places the drifts piled up eighteen or twenty inches, according to the *Danville North Star*. The new shoots on the branches of trees froze. So did all the corn and garden vegetables just out of the ground, and many of the local sheep and cattle. A farmer went out of his house to look after his sheep in the midst of the storm. "If I'm not back in an hour," he jokingly told his wife, "call the neighbors and start them after me. June is a bad month to get buried in the snow." Three days later searchers found him, frozen to death.

In this year that would be long remembered in New England, summer never came. Instead, the sun rose each morning red and rayless as in a cloud of smoke, and so many birds died in the July ice storms that for years afterward the spring was almost silent. The cause of these freak atmospheric events is now believed to have been a combination of sunspots and the globe-encircling cloud of dust from the eruption of Tamboro in the East Indies in April of the preceding year.

Whatever the explanation, the loss of an entire season is a rare but harsh reminder that we can do little about cold weather except light a fire, put on an overcoat and adopt the stoic attitude of an old New Englander. One such New Englander was Robert Frost, who had some thoughts about the winter killing of a tree that he had planted:

"It is very far north, we admit, to have brought the peach . . .
Why is (man's) nature forever so hard to teach?"

Boundless as we might like our horizons to be, there are certain cosmic events that govern our lives. As the earth spins, day turns into night, and during 365 succeeding days and nights the earth completes one elliptical orbit around our energy source, the sun. Because of the earth's permanent tilt during orbit, we have seasonal changes in daylight and temperature. For half a year at a time the Northern Hemisphere cools under the slanted rays of a remote sun, while the Southern Hemisphere swelters under a direct, hot sun. Then the process is reversed.

Only a small portion of the sun's rays reaches the earth after a 93-million-mile journey through space, but these rays are effectively trapped in the earth's atmosphere. Absorbed by certain leaf pigments of plants and by water vapor in the air, radiant energy from the sun is at once the basis of life and the impersonal force that powers the circulation of the atmosphere, spawning our weather systems. It causes evaporation that makes the rain and the snow, and stirs up cyclones and hurricanes. Moisture is dispelled on the trade winds to the middle latitudes, and to the polar regions, where legions of snow clouds gather each winter for an assault on the Temperate Zones.

At this time of year North America is a natural battleground between the arctic and the tropics. Polar "fronts" linger on the Canadian border, advancing in cold wave after cold wave down the Great Plains and the Mississippi Valley to meet token forces of moist, tropical air from the Caribbean. The clash is often explosive, setting off a line of eastward-moving snowstorms.

Then, with the arrival of summer, the North warms, and the climate of the continent equalizes. Clear skies prevail, and strong tropical heat waves invade Canada. The storms of this season are mostly local downpours, caused by rising convection currents from the heated ground, carrying aloft to cold, upper altitudes the vapor given off by lakes, streams and vegetation.

Regardless of season, high mountains make their own weather. The Front Range of the Rockies rises abruptly out of the Great Plains like a granite fortress against the Canadian blizzards that move south and east. Sometimes southwest winds called "chinooks" blow off the mountain slopes to melt winter snowbanks in the foothills overnight, as in a spring thaw. Farther west, the Coast Ranges intercept the fog and rain drifting in from the Pacific, to cast spectacular rain shadows. The wettest spot on the continent, with over 140 inches of rainfall, is the windward side of the Olympic Mountains of Washington; the driest is the desert area in the lee of the Sierra Nevada, where there is less than five inches of rain every year.

Within the broad outlines of climatic zones, the surfaces of land, water and plants radiate heat, lose moisture to evaporation and deflect wind at different rates, creating microclimates that may be either wetter or drier, hotter or colder, than the region as a whole. And there are other, very specialized climatic influences. An ant heap is a mountain in miniature, within which the ants regulate the temperature of their eggs by moving them from side to side as the intensity of solar radiation varies on the outer slopes.

To help them cope with their environment, many Temperate Zone plants and animals have evolved ways of adjusting to the seasons. After the retreat of the Pleistocene glaciers, large areas in the North could not have been colonized if living organisms had not developed natural rhythms governing reproduction, molt, hibernation, and

migration. These rhythms are tied to cosmic light periods rather than to less dependable temperature changes.

The biological clock seems to be wound by increasing periods of light. In Southern California, as spring comes on, a white-crowned sparrow is conditioned by 14 more minutes of daylight a week to develop gonads and fatten up for a trip that will culminate ten weeks later on summer nesting grounds in Alaska. Light is the only reliable source of measurement by which the bird knows of the approach of long hours of daylight and an impending abundance of food at its summer home. Light also governs to a remarkable degree the cell division, growth, turgor pressure, spore discharge, and luminescence of many kinds of plants. To move them away from their established orientation in time and space can be disastrous. Grama grasses that flower in Arizona when the day is twelve hours long will fail or flower late if the period of apparent daylight is artificially increased to sixteen hours. These species cannot be grown on the northern Great Plains, where summer days are long.

Most responsive to temperature are cold-blooded animals. The fry of some photoperiodic fishes that normally spawn in the spring may turn up during a warm fall, along with a scattering of spring flowers. Their biological clocks have been upset by the unusual heat.

After some 200 years of research, we know of no pigment that could be a photoreceptor, nor of any specific mechanism to explain the phenomenon of the biological clock. Yet there is no doubt that it exists. Photoperiod affects the embryonic leaf as it does the mature leaf. It is registered by some seeds in the ground and by larvae in cocoons. In animals it is perceived by the eyes and, in some unknown manner, by the pineal gland, the nonfunctional "third eye" first discovered in the head of a lizard of New Zealand, the tuatara. There are similar structures in the center of the skulls of all reptiles and of many fish, birds, and mammals, including humans. The pineal gland manufactures a hormone that suppresses sexual development either directly or indirectly in response to light. Photocells implanted near the site of the pineal gland in the heads of sheep have registered altered readings, depending on whether the animals were standing in sunlight or in shade. The function of the gland seems to be impaired if nerves connecting the eye to the brain are severed; its rhythm is dampened, as though by continuous darkness. There is still much to be learned about the complex and endlessly varied natural rhythms affecting the behavior of living organisms as the seasons change.

Across the continent, March and April are the windiest months, and the atmosphere is charged with electrical storms; winter is weakening but still holding on, while the South rapidly warms. Spring comes to Florida four to six weeks earlier than to New England, with an advance guard of low-flying clouds and drenching but brief thundershowers. The season arrives with the intense green of new cypress needles, a galaxy of minute green flowers in gray streamers of Spanish moss, and the chattering of migrant birds passing through from the West Indies and South America.

Here no lone robin suddenly appears. Hundreds have filled the cedar and myrtle trees all winter, but now—as a prelude to their departure—they descend to the ground to scratch for worms. Migrating North, the first individuals appear on Northern breeding grounds just as the sun's rays honeycomb and dispel the snowbanks and cause the ice in the rivers to crack with resounding booms. The melt-off of snow softens the ground, mixes in the dross of last year's plant growth, and brings up earthworms. The change from storm to calm—the impact of the sights, sounds and smells of spring—are more sudden and memorable in the North; but the return of life to the landscape, of the insect and the worm before the fledgling bird (one being necessary to the other) is universal.

Like the white-crowned sparrow in Southern California, the robin reacts inwardly to lengthening days on its Florida wintering ground. Triggered by increasing light, the hypothalamus, in the posterior part of the forebrain, sets off a series of internal adjustments that ultimately result in the release of gonadotropic hormones into the

On Winds of March
Riding out a spring squall, a turkey vulture soars high above earth. The great bird can plane on turbulent air for hours with scarcely a flap of its broad wings.

Insect Pollinator
*Crawling over a milkweed flower on the
Mojave Desert in search of nectar,
a wild bee steps into masses of sticky
pollen that will cling to the bee's
hair-covered legs. When the insect moves on
in its continued hunting for food, the
pollen carried along will serve to
fertilize other milkweed flowers visited.*

bloodstream. These hormones stimulate the production of sperm cells and bring about a change in song and other behavior. But what signal of approaching spring is used by the hundreds of migrants that winter farther south, beyond the West Indies and Mexico? Are the slight changes in day-length within ten degrees of the Equator in northern South America enough to tell warblers the time of year? Or are the warblers conditioned by weeks of unchanging twelve-hour days? Do bobolinks wintering on the Argentine Chaco respond to *shorter* days after March 21st, which is the fall equinox in the Southern Hemisphere?

Whatever causes them to migrate, the birds seem to be drawn northward by increasing temperature. On the West Coast, which is warmed by ocean currents in the winter, some of the early spring migrants encounter an average temperature of 35 degrees F. all the way to Alaska, but others, in the Dakotas, may be stopped by deep snow and freezing temperatures for a month or

more. The sudden melt-off of snow between mid-March and mid-April speeds the Northern birds on to their Canadian nesting grounds and opens up the prairie ponds for resident waterfowl.

Evidence of the irreversible character of springtime behavior is provided by birds in years of unseasonable weather. In May of 1939, for example, hundreds of scaups, surf scoters and eider ducks were held up by the cold and could not reach Canada in time for the nesting season. They were seen laying eggs in gulls' nests on the islands off Jonesport and in Boothbay Harbor, Maine.

Compared to the horizontal progress of the spring season at sea level, moving at an average 15 miles per day towards the arctic, spring creeps up mountainsides at an average rate of only about 100 feet a day. A few yards in altitude are the equivalent of hundreds of miles in latitude. The green tide of spring floods the valleys and twists a serpentine path through mountain gaps long before invading the thinner and colder realm of the upper heights. Where juncos and pipits nest

in the high tundra, spring is timed with the awakening of lands within the Arctic Circle.

The first spring arrivals in the Northeast are small flocks of wild, wary males. The robin of New England is not yet in song because his mate is still some weeks of flying time away. But the male bluebird precedes his mate by only a matter of days. He quickly finds a home in a hollow tree or fence post and trills until she appears. With a flaring of brilliant blue feathers, he caresses her with his beak and offers food as an inducement for her to follow him to the nesting place. If another male chooses to contest for the same mate, the ceremony turns into a brawl. All of the birds roll in the grass, rough and tumble, and any female who appears to have had a prior claim on the invader may join the fracas.

As noisy harbingers of spring, the bluebirds vie with song sparrows just coming out of their winter thickets, and with waves of blackbirds passing through the marshes in dusky, clattering crowds. The red-wings that nest in the marsh make their presence known by very singular behavior on the part of the females. Unlike the secretive individual of March, the homemaker of April aggressively invites the male red-wing to pursue her in a fast, spiraling chase.

Wildflowers bloom in this early spring landscape before the sedges green up and the buds pop out on the trees to hide the nests of incubating birds. Who has not walked in snowy woods and found the crimson-cowled skunk cabbage poking its dark head out to greet the sun? The leaves of the plant actually absorb and radiate enough heat to melt the surrounding snow. With longer hours of sunlight, hepaticas, wood anemones and bloodroots emerge from their warm beds of dead leaves. (Sunshine, slanting through the dormant branches above, heats the leaf litter of the forest floor to such an extent that the buds of flowers buried within are as much as 10 degrees warmer than the world outside.)

The chirp of the black cricket tells us that in-

27

sects will soon be appearing in all their myriad forms—the wild bees and flies from pupae in the soil and litter, grasshoppers from eggs, leaf hoppers and mourning cloaks from their beds of hibernation under stones and bark, and new butterflies from chrysalids.

Having been stored away for the winter, an insect merely resumes its life cycle wherever that cycle left off. None grows to maturity without the shedding of many skins between egg and adult, and an extra stage of existence—in the pupa of the fly, the cocoon of the moth, or the chrysalis of the butterfly—is necessary for the large number that undergo complete metamorphosis.

Within the chrysalis the processes of life and death are merged. The visible "mummy" is only a façade behind which takes place the larva's dissolution and reconstruction into a slender insect with legs, wings and compound eyes. There is no trace of the crawling, leaf-eating caterpillar of the previous season in the butterfly that now launches forth on its winged search for pollen and nectar.

Conditions of temperature and light when butterfly eggs are laid or shortly thereafter bring about the release of hormones that later cause "diapause," or arrested development. Without such a check, fall insects would emerge and die in the winter; instead, they come out in the spring. The second or third hatches of eggs during the year are not inhibited in the same manner, and the chrysalis stage for these latecomers takes as little as ten days to complete. The difference in length of time spent pupating causes striking variations in the characteristics of different generations. The zebra swallowtail of the spring is smaller and paler, and the pearl crescent is darker than any summer individual of the same species.

Unlike the birds, migratory butterflies such as the painted lady and the monarch mate on wintering grounds in Florida and California and lay their eggs on the way home. Two or three months —and as many generations—later, they have spread to the northernmost part of their range in the United States and Canada. Here the last generation of the year develops, growing into the adults that gather in great bands and stream south in the fall.

Having evolved together in the geological springtime of the earth, flower and insect are closely adapted to each other's needs. Although the dandelion and the violet of the early spring season need no insect pollination, most of the showy flowers of May and June do. So they are shaped, scented and colored to attract the tiny creatures whose hairy bodies distribute fertilizing pollen from one flower to another.

The honey bee sees only the blue range of the spectrum and the ultraviolet (invisible to us) which is reflected by some red flowers. The bee associates the color and shape of a flower such as the blue flag iris with the nectar that lies at the bottom of the passageway formed by a long corolla. To enter this passageway, the insect must brush against the stigma, leaving behind a considerable amount of previously collected pollen; its body is redusted, on the way out, by the anther. Once rewarded in its search for nectar, the bee will return day after day during the hours of peak nectar production. It navigates by the pattern of polarized light in the sky. Amazingly, it can pass on information about the location of a flower bed to other bees by performing an elaborate dance within the hive.

To the hunting spiders, and to the wasps that prey upon them, flowers are places to lie in ambush. The ground below is populated by many other miniature predators—including the ant lion larva, waiting with raised pincers at the bottom of a pit for an ant victim to tumble in, and the ground beetle lurking under a stone.

Nowhere does the presence or absence of moisture have a more dramatic effect on the timing of spring flowers and insects than in the Southwest, where unrelieved drought alternates with downpours that etch deep arroyos and fantastic badlands across the landscape. If no rain falls in the Mojave desert, the pupae of bees, beetles and moths and the seeds of flowers stay asleep in the dry, hard soil, and the Joshua trees fail to bloom, sometimes for years. The record low rainfall for the region was one hundredth of an inch, all of which accumulated during one month (March,

Jumping Cholla
The small cholla cactus of the Southwestern desert has tips that break off easily, and are often carried by the wind in "jumps" to start a new cactus colony far away.

The Noisy Bullfrog
To summon females, a bullfrog sounds its loud bass with mouth closed. Living much of the time out of water, the frog has developed both a vocal sac and a large ear.

1919) in a three-year period between 1917 and 1920. The maximum rainfall that can be expected in the comparatively lush Arizona upland desert is about eight or ten inches in a year.

In the deserts, perennials like cactus and yucca form stands along drainage channels where they become more or less entrenched, regardless of the climatic fortunes of the year. But annuals need regular watering to grow. Their seeds are coated with chemical inhibitors that must be washed off. These rain gauges are adjusted so that many seeds do not germinate, even under the right temperature conditions, unless there is at least an inch of rain. Once stimulated, the long-stored energy of the plants is spent in a sudden burst of flowering, after which they seed and wither away. Beneath the cholla and saguaro cactus of Arizona, the desert floor is carpeted with a rainbow array of spring annuals from late February to mid-April. Summer showers, if they come, trigger another display of annuals—desert flowers of those species that will germinate only in hot weather.

Mammals are aroused in the spring from all degrees of winter sleep—tree squirrels and bears from dormancy, prairie dogs and ground squirrels from hibernation. The deepest sleepers lie in a condition of cold storage close to death, from which they can rarely be awakened before their biological alarm clock goes off (although laboratory tests show that at least one—the woodchuck —can hear sounds while it is cold and immobile).

When curled in hibernation, the woodchuck has a heart beat of 4 pulses a minute and a rectal temperature of 38 degrees F. Just four hours after the wake-up signal is received; the endocrine glands have taken over, stores of thyroxin have brought the heart rate up to 75 a minute, and the temperature of the animal rises to 98 degrees F.

The ability of hibernating animals to generate

The Silent Salamander
After a rain, a tiger salamander emerges from its pond and crawls up the muddy rim to catch
a worm. Unlike the frog, this amphibian makes no sound and hears nothing.

heat in a hurry, whenever they awaken, explains how a woodchuck may be able to venture out in the snow as early as February 2, "Ground Hog Day." Such precocious individuals go back to sleep, whether they see their shadows or not, because there is little food available at this season in the Temperate Zone. Most woodchucks are more sensibly programmed to appear in March, when their tracks, leading from hole to hole, tell of the annual search for a mate. The young are born a month later, and at the age of eight weeks begin to fatten up on tender meadow plants.

Many mammals that breed in the fall—bats, martens, fishers, badgers, weasels and armadillos included—produce spring babies, due to another special internal mechanism. Delayed implantation of the fertilized egg postpones the development of offspring for long periods. Although the resourceful mother badger may be active enough in the winter cold to dig out a hibernating rodent

or snake for a meal, the seeds of new life lie dormant within her. There is a delay of some two months before the fertilized eggs are implanted in the uterine wall. The embryos only begin to form in February. The striped-nosed badger babies are delivered in a grass-lined burrow in April, May, or early June, and are weaned in July—at a time to be raised sumptuously on young ground squirrels or prairie dogs.

Reptiles first become active during the lengthening days of spring at an average temperature of 50 degrees F., and begin sunning themselves in the open, but amphibians wake only to hide in shaded places where the coolness and moisture protect their thin, permeable skins. Spring rains prompt hordes of frogs, salamanders and toads to migrate, under cover of darkness, to newly-created ponds for egg laying.

The trill of the spring peeper is an indicator of

the season, from the Northeast to Florida and Texas. One singer encourages another until a high, metallic chorus of sound reverberates from the inflated vocal sacs of dozens of tiny males, floating or clinging to aquatic vegetation. In Southern ponds the sound is underlaid by the harsh strumming of the cricket frogs, and an entire orchestration of bachtrian calls, including the bull frog's deep hr-r-rmph, rises to a tremendous din.

The ponds are thick with amphibians' eggs as early as February in Florida and up until July in Maine. Hatching in less than a week, if the air temperature is as high as 72 degrees F., the gill-breathing larvae grow and normally transform into tiny froglets, toads and salamanders by the end of the summer. Among certain exceptions are "tailed" frogs of the cold mountain streams of the Pacific Northwest that transform a year after hatching, and Western tiger salamanders in iodine-poor waters that fail to transform. These freak salamanders breed in the gilled stage. Where the shallower ponds dry up early in the season, they are doomed to die.

Dependent on water as an environment for breeding, the amphibians are closer to fishes than to reptiles. As for the fishes, these "cold-blooded" organisms prove extremely sensitive to the slightest variations in water temperature in rivers and lakes; some can detect a shift of only a fraction of a degree. As the warm spring air heats the surface, a level of sudden temperature change, or thermocline, is established, and at this level the whitefish stay. Above them the sunfish, perch and bass search the surface for insects. Catfish and bullhead feed on plants and crustaceans in the muddy shallows, while the brook trout prefers the coldest water at the deepest bottoms. In winter, bottom feeders range up to the thermocline and surface feeders become sluggish, often burying themselves in the mud.

The spring-fed rivers of northern Florida, having constant temperatures the year round, and very little variation over the decades, are perfect natural laboratories for observing the effect of day

Golden Shiners
Minnows school in the sunlight that penetrates the clear waters of Weeki Wachee Springs, Florida. Beds of eel grass, covered with algae, supply food for these fish.

length on the breeding of fishes. The Weeki Wachee River, as an example, constantly registers 74.2 degrees F. Here no sudden hot spell can be responsible for initiating the springtime renewal of life under water: the heavy growth of algae on the eel grass, the bright colors and the musky odor of fishes laying eggs on the bottom in circular "beds." It is a measurably longer period of daylight.

The cold water fishes of northern climates apparently need the opposite clue; the brook trout, an autumn-spawning fish, matures a month earlier than normal under artificially shortened days.

Salt water fishes such as salmon and shad find their way back to the rivers where they were spawned at their own spawning season by many indicators which have not been fully sorted out by marine biologists—temperature, currents, degree of salinity and organic odors in the water. Young fish have better recall of learned situations than older ones, and the farther they go from "home" the more accurately they seem to remember the way back. They can swim the seas for three or four years and then return unerringly to mate in their original streams. One of the early-flowering

shrubs of New England was named the shadbush by people of Colonial days for the April "run" of the shad in the Eastern rivers.

In the rookeries of the wading birds, no eggs hatch before the minnows swim the rivers in schools, feeding on new plant life. The largest of the rookeries on the Weeki Wachee is a five-acre black willow stand in the midst of a sinkhole lake. In this highly protected woods—actually a swamp surrounded by a moat—snowy egrets, American egrets, anhingas, little blue herons, green herons, Louisiana herons, cattle egrets, red-winged blackbirds and boat-tailed grackles nest together and proclaim the fact vociferously. The heron pairs spell each other on flights to hunting grounds along the river, for they must defend their eggs and hatchlings against the poking bills of careless neighbors and the thieving habits of fish crows. These predators will seize any opportunity to rob the herons of their first, second, third and even fourth brood of the season. It is largely the persistence of the crows that keeps rookery production lines running far into the summer. Beneath the bird city, numerous water snakes and alli-

34

Coming Out of the Shell
Using an egg tooth still visible at the tip of its bill, a Louisiana heron (opposite page) struggles to enter the world. From first pip to final breaking out, the process has taken the bird about twelve hours.

Hiding in a Hollow
Not until the bluebird is about two weeks old, and feathered out, does it see the outside world. Then the nestling watches for its parents at the doorway to its treetrunk home, begging for food with bright orange gape.

gators gather, ready to snap up any fish that may be dropped or any young that may fall from the nest.

A striking antithesis to the loud, sociable heron is the Eastern bluebird, whose young grow up quietly in a hole in a tree. Among the earliest species to nest in the North, bluebirds raise two and sometimes three broods a year. As soon as the first-brooded young have pushed their way out into the world, the male takes charge, teaching the young birds how to fend for themselves.

On June 22, the summer solstice, the sun shines down with greatest intensity in the Northern Hemisphere, beginning a season of siesta, of rest and growth. In the subtropical Florida Everglades sweet water flows, bringing to an end the long dry period during which moisture was to be found only in alligator ponds and in the brimming, cup-like bromeliads that festoon the cypress hammocks. By contrast, the Southwestern deserts are now undergoing an intensity of dry heat from which animals escape by digging down into the earth. When forced to move on the surface, they skim quickly across the hot sands—lizards on

fringed toes, sidewinders on loosely thrown coils.

Over much of the continent the days start with clear skies and rising winds, and may end with thundershowers. In the scorching interlude between the cessation of the wind at about one p.m. and the lengthening of shadows, scarcely a bird flies, mammal runs, or reptile crawls. But various insects, protected by waxy or chitinous exteriors and pale, heat-reflecting colors, are at their peak of activity at the height of noon. Heat radiation dissipates quickly at ground level, so the ant and the grasshopper keep cool by walking on tiptoes, stretching into a different microclimate just millimeters above the earth.

Late summer is insect time. Cicadas and katydids fill the air with the sound of their shuffling and scratching, which is greatly intensified by heat, as is the stridulating of crickets. The cricket sounds work up from rusty croakings to persistent chirping so regular that one can nearly always work out the temperature within a couple of degrees by counting the number of chirps made by one insect in 14 seconds and then adding 40.

Spiders barely noticeable in the spring are bigger now and spin more gossamer, filling every

Sidewinder
Across the hot surface of the Sonoran Desert, a sidewinder rattlesnake travels swiftly with its characteristic undulating, sidelong movement, leaving a series of parallel scratches in the sand.

A summer sign is the empty shell left by a nymph which has transformed into a dragonfly.

available nook with their great sheet webs and expansive orbs. Their discarded draglines, delicate strands once floated for the purpose of swinging across gulfs between tree and tree, collect on the meadows like cobwebs blown from a dusty house.

After darkness has settled, the cold blue lights of fireflies flash. The firefly's fuel, luciferin, is oxidized under control of the insect's nervous system, and warmer evenings induce spectacular displays between females hidden in the grass and males flying overhead.

Dragonfly nymphs usually transform at night, when birds and fishes are not likely to seize the helpless new insect, which must cling for several hours to a water weed beside its discarded skin. In a period of only a few minutes, the misshapen abdomen pushes out to full length, color suffuses the pallid skin, and two pairs of crumpled wings unfold. Then the insect gradually adjusts to its new state. As sunrise touches the pond with light, the dragonfly takes to the air. No other creature of summer is so supremely diurnal and aerial as the big darner—hunting, mating, and often laying eggs while on the wing (the female dropping her eggs onto the surface of the water).

In this season of productivity, seeds lie in windrows in the gulleys, and the young of many mammals and birds are out of their nests and burrows. But less food is being consumed than at any other time of the year because so little energy is needed to maintain normal body temperature. According to the carefully compiled notes of Frank and John Craighead, hawks and owls eat three to eight per cent less in summer than they do in winter, de-

pending on body weight. The larger the bird, the less food it must have, proportionately, to keep up its energy reserves. The same holds true for mammals. The large body of a deer both heats and cools more slowly than the small body of a mouse; it uses food more efficiently.

Biologists once observed a herd of western mule deer through all the seasons, and found that their activity was closely related to air temperature. Most of the herd could be seen in one day in the spring, but only about half appeared in the summer, after they had left the dry open hillsides to browse in the woods and canyon bottoms—the bucks in velvet, the does with their new fawns, the yearlings in small groups, all scattering to feed on shrubs and wildflowers. As early as April, the deer avoided clearings on unseasonably hot days of 78 degrees or more. As summer advanced, they were less and less active during the day. When the late morning temperature reached 88 degrees in June, they made a habit of bedding down until three, and did not fully resume feeding until about six o'clock in the evening. By July, few or none would venture into the open canyons until after dark, some going directly to water while others browsed steadily through the night.

High up in the Rockies, the Sierras and the Cascades, the alpine meadows are just warming and flowering in mid-July. On the exposed slopes, snow fields melt into bogs. Here wapiti feed on sedges and grass in the daytime, in a climate appropriate to May, and bed down at night in clumps of wind timber. In six weeks the brief summertime awakening will be over, the wapiti will have migrated to lower elevations, and the mountain top

will once again be locked in a white stillness.

At plains level, the forest edge is a midsummer retreat for the hawk and the deer; the shrubs provide cooling for the jackrabbit, the marshes and bogs for the meadow vole, an underground burrow for a ground squirrel. At noon of a summer's day the prairie may have an air temperature of over 98 degrees F.; at the same time the shrubs will register 87 degrees F., and the burr oak forest 77 degrees F. (The corresponding soil temperatures for these environments would be 79 degrees, 74 degrees, and 70 degrees, respectively.)

Without the vigorous rains of spring, the prairie potholes and sloughs begin to dry out, dispossessing the waterfowl. Clouds pile up in the summer sky, but to no avail. Flying at an altitude of 5,000 feet, the mallard duck will have wet feathers, but at 1,000 feet she is perfectly dry; the long drops of rain evaporate well before they reach the earth. Luckily her ducklings are ready to leave the nest now, and the drake has already migrated to some larger lake or river—perhaps as far away as Klamath Basin, California—to wait out an annual molt initiated by the photoperiod in late June and July. The wing feathers are dropped, all in a matter of hours, and he must hide in an "eclipse" plumage as dull as the female's until new wing quills grow and he assumes his bright winter plumage of green and russet.

The few large mammals that can stand the heat of the plains cool their bodies by sweating or panting, and by adopting light summer fur. The fluffy winter coat of the coyote is shed for one so slight that the animal resembles a small fox—and is often mistaken for one. The pronghorn's ankles grow bare, exposing the outlines of strong tendons. Its thin coat can be raised at will by stiff guard hairs on the back, to ventilate the skin while running. The bison becomes a scurvy-looking beast, with hanks of matted hair clinging to its shoulders and much of the softer fur on its sides and hindquarters worn off.

The changing of fur is a substitute for major internal adjustments to changes in climate. If

The Cool Coyote
A coyote loses most of its fur in the summer, leaving only the outer guard hairs. This gives the animal a ragged appearance, but adds greatly to its comfort.

subjected to average winter temperatures in its summer coat, for example, the arctic fox would need to increase its food and oxygen intake by sixty per cent to function normally. Even burrowing animals need to change their fur coats twice a year—the winter fur of the mole being 11 to 12 millimeters, and the summer fur only 9 to 10 millimeters thick. This slight difference makes the mole more comfortable when it tunnels close to the surface in summer. It will go down three or four feet in the winter time.

Very small animals do not have sweat glands, as their bodies might quickly be dehydrated through sweating. For the masked shrew, weighing no more than four grams, the only recourse in hot weather is to find a cool, moist place to be. Since the metabolic rate of an animal varies in inverse proportion to its size, the shrew needs to eat almost constantly, winter and summer. Shrews as a whole are believed to represent the lower limits of possibility, as far as size is concerned, in mammals. The daily rhythm of one of the biggest, the water shrew, is about thirty minutes of activity followed by sixty minutes of rest, day and night. There are two major peaks of activity, one just before sunrise and one between sunset and midnight, periods when the air temperature is low and the animal's cold-blooded prey is easy to catch. The shrew matures in a few days, and lives no longer than a year and a half. Some shrews are able to immobilize insects with a poisonous saliva. Often they store food for future needs.

Less highly geared metabolically, mice and voles are nevertheless nervous, active creatures that store food at all seasons. Their larders reflect an abundance of a favorite dish, a gourmet preference rather than a strict necessity. Look in a hollow stump in a bluegrass meadow and you may find about a gallon of the yellow fruit of the horse nettle, tucked away there by the prairie vole, whose regular fare is more often roots and tubers.

The squirrels and chipmunks that seem to delight in the heat of summer in the humid East have made different adaptations to the arid West. The Eastern chipmunk spends the hottest weeks of July below ground bringing forth a second brood, because there is still plenty of food. But on the inner ridges of coastal mountains in the West, chipmunks are curled up with their stores of seeds in summer torpor, or estivation. This is their way of weathering the long dry period from June to September.

Here, and in the Mojave, most of the ground squirrels fatten in June and then go to sleep for eight months, a period corresponding to both estivation and winter hibernation. There appears to be no difference between these conditions. As the animal becomes torpid, in about six hours, its temperature drops close to that of its cool burrow, and its breathing and heartbeat become barely perceptible. Its sole source of energy now is body fat, even though provision has been made by the animal for eating.

Coasts are normally the coolest parts of North America in summer. Winds off the Labrador Current bring welcome relief from heat waves to the Atlantic Shore. On the other side of the continent, heat from California's interior valley is tempered by sea breezes, to keep the Pacific Coast mild and foggy until fall. Surprisingly to Easterners, September is the hottest month in San Francisco, even though the sun has withdrawn to the Equator.

At the vernal equinox, September 21, trade winds in the Northern and Southern Hemispheres shift direction, and storms of hurricane proportions develop with increasing frequency. In the "cradle zone" southeast of the Windward and Leeward Islands about four out of ten tropical storms mature, to make their slow and often destructive journeys up the Atlantic Coast or through the Mississippi lowlands in September and October.

The approach of a hurricane is signalled by showery, squally weather for twenty-four hours or so, during which high, fluffy clouds speed by. The barometer falls. Birds practicing their migratory formations scatter, and armies of rodents evacuate the open fields for the shelter of the woods. They sense the gathering storm even before the sky turns an eerie yellow and the sun is obscured by stratified cloud layers. The dark eye of the storm passes, with the fury of thunder, heavy rain, and winds of 75 miles per hour or more, and the hurricane blows itself out over land or sea.

Autumn brings the quiet warmth of Indian summer days, dusty with dry leaves and hazy with the smoke of brush fires, and the surprising chill of lengthening, frost-bearing nights. Deciduous trees respond by shutting down their interior water

Autumn Woods

"A light haze rests upon the moving landscape, the many-colored woods seem wrapped in the thin drapery of a veil; the air is mild and calm as that of early June . . ." Francis Parkman might have had in mind a scene very much like this shadowy New England glade when he wrote the foregoing lines.

works and cutting loose flotillas of leaves. The colorful drifts pile up on the ground in an ordered sequence; in New England the glowing red of soft maple is often followed by the purple and gold of ash and sugar maple, and, in a last breathtaking display, by the scarlet-leafed oak. The yellow pigments have been in the leaves all summer, and are merely unmasked when the green drains away, but the red pigments emerge as by-products of chemical breakdown. They are especially showy if bright sunlight accompanies the short days of autumn. With the approach of cold weather, the vines, shrubs, and trees use various sugars to help block the coagulating effect of freezing temperatures on sap. It is partly the dissolution of these sugars, trapped in dying leaves, that bronzes the landscape. If the chemical reactions are interrupted—say, by an early freeze in August—leaves may blacken or fall off while still green.

On the wings of cold fronts, migratory birds now stream south, their major flight lanes converging along river valleys, coasts and ranges as the continent narrows. Preconditioned by glandular changes and fat-storing, the birds may travel thousands of miles to southern wintering places, navigating by the stars and the sun. The young fledged before these journeys seem to have an instinctive sense of direction; they precede the adults rather than following them. The shortening of the day is the strongest stimulus to migration, especially in the far North, which is deserted before the food supply becomes scarce. Beginning with shorebirds, swallows and nighthawks in August, the birds on the wing build up in numbers through September, reach a peak in October, and diminish to a few stragglers by late November.

The leaden skies and southerly winds of late summer give way to roaring winds out of the

Preparing for Winter

Getting ready for the onset of winter, the prairie dog (above) has only to eat substantially and store up fat in preparation for months of hibernation. The pocket gopher (below) does not hibernate; instead, this rodent crams food into huge cheek pockets and carries it to underground bins from which the animal will be able to feed at leisure during the cold weeks ahead.

northwest as the weather clears and a cold front moves through. This is the time when hundreds of hawks and eagles from all over New England and above the St. Lawrence in Canada take advantage of the daytime updrafts along the ridges of the Connecticut Valley, soar past Mt. Tom in eastern Massachusetts, across the Hudson River near Bear Mountain Bridge, and over Hawk Mountain, a high and slender point on the Kittitinny Ridge in Pennsylvania, as they make their way south. On the hillsides below, song birds swarm by the thousands, feeding and resting. Night is their time to fly.

While some mammals fatten under short days, others store food more frantically than ever. The Eastern chipmunk crams its underground vaults with more than a bushel of nuts, and the little bean mouse of the plains, which the Sioux call *Tunka*, hoards five or six bushels of wild beans, roots and tubers. The squaws used to dig up these vegetable stores, which were an important part of the Indians' diet in winter. In return, they sometimes left corn or other food in the empty caches.

Among deer, the bucks appear in force for the rut as the days begin to grow shorter. Their antlers harden and shed the velvety covering. Other animals are busy mating. The marbled salamander breeds at this time of year, the female depositing her eggs in a depression in the forest floor. If it rains and a puddle forms, the eggs hatch; the larvae will transform the following summer. Otherwise the eggs lie dormant until the coming of spring rains.

Males of fall spiders tug insistently on the outer strands of webs to call females. After the mating and depositing of the eggs, nearly all the adults die. Their silken egg sacs remain, some glued to stones, others tied to twigs or folded in leaves. The new generation stays in the egg sacs, waiting for the proper spring temperature before breaking out and ballooning away.

Many short-horned grasshoppers and June bugs copulate in the early fall and plant their eggs like seeds. The female grasshopper drives her sharp-tipped abdomen deep into the earth and then spreads two prongs wide enough to excavate a hole for the eggs. After depositing the eggs, she sprays them with a bubbly liquid that hardens to become sponge-like, ensuring insulation and aeration. Usually they will not hatch without passing through a cold period first, a necessary mechanism to prevent the emergence of larvae during a mild spell in mid-winter.

Plants, too, are protected by inhibitors. The seed of the birch and the microscopic leaf buds of most trees are formed almost a year in advance, under the influence of long hours of daylight; they fail to imbibe water or to grow until they have first undergone a period of cold and then felt the return of the long days.

During Indian summer nights, frost chills the soil more and more, week by week. A temperature inversion occurs not unlike the inversion to be found in rivers and lakes, whose waters are warmer in winter at greater depths. The "frost line" in the soil is the equivalent of the thermocline in water. Mites and microarthropods migrate down in summer to keep moist; they become immobile when frozen, and active again when thawed. Other life in the teeming soil, however, cannot stand freezing. The mole, the shrew, and the hibernating worms and insects migrate well below the frost line. The extent of their journeys depends on the depth of the freeze, which increases with latitude, and upon the sensitivity of the animal.

On the Kansas prairie, where there is little snow to provide insulation, dung beetles dig winter burrows 31 inches below the surface, twice as deep as their summer egg burrows. In Minnesota the soft larva of the May beetle, or "white grub," burrows two feet down, but the tougher larva of the click beetle, or "wireworm," winters only five to seven inches from the surface. As little as half an inch of snow cover provides an important margin of protection in sub-zero weather. Nevertheless, a sudden winter thaw and freeze may completely upset the temperature balance below earth, killing many of the soil's inhabitants.

As winter approaches, animals disappear from their accustomed haunts one by one. Hibernating snakes den up in tangled heaps in old logs; salamanders crawl into rock niches; frogs and toads dig themselves into solitary cells in the soil, below the frost line. The hibernating woodchuck, too, digs as deep as necessary to avoid freezing, while drowsy flying squirrels huddle in hollow trees by the dozen. Tree bats migrate, and cave bats extend their daily periods of torpor over an entire

A Late Mating
At summer's end, a pair of giant Florida lubber grasshoppers instinctively mate before cold weather comes. The female will bury her fertilized eggs in the ground and cover them with insulation to withstand the cold until it is time for them to hatch out in the spring.

season, stirring in their dank environment of about 70 degrees F. only to drink water.

At this time of year the gonads of hibernators are inactive. Energy reserves are high, and so is the store of glycogen in the liver, which acts as a general anesthetic. Meanwhile the hypothalamus in the brain takes over and, through the nervous system, adjusts the blood flow, heart rate and breathing. No one knows exactly how to explain this change away from a glandular control of body processes by the endocrines, which are now immobilized so that the animal is completely incapable of reproduction or growth.

For mammals, the inversion of bodily processes is a gradual one. The woodchuck comes out less and less frequently and sleeps longer and longer in its den, as the nights lengthen. The animal's daily withdrawal, coinciding with the rhythm of darkness and light, helps to set its biological clock.

Once sound asleep, the woodchuck maintains a body temperature slightly warmer than the nest. Its metabolic rate wavers between 1/30 and 1/100 of the normal rate during sleep, except in emergencies. It can double or even triple the respiratory rate, without waking, to compensate for a drop in the surrounding air temperature below the critical point of 35 degrees F. If the cold snap continues for longer than a day or two, the animal wakes but does not necessarily eat; it depends instead on stores of fat, including a gland between the shoulder blades—the so-called "brown fat"—which supplies instant energy in the form of carbohydrates through the impetus of the nervous system. Such awakenings take up enough energy to account for the loss of one third to one half of a woodchuck's weight during four or five months of hibernation.

By contrast, the dormancy of the bear, the skunk, the badger and the raccoon is irregular and depends on the weather. Only the females sleep for any extended period. And their sleep is not very deep. The female bear has a temperature of 61 to 66 degrees F. when she gives birth to young in the middle of winter, and she suckles them in this condition.

For the nocturnal predators abroad in winter —mountain lion, fox, lynx, coyote, and owl—the long hours of darkness mean more time to hunt in a lean season when demands on their energy are most severe.

Some of the hunters and hunted have made very definite adaptations to stalk and avoid being stalked in a white world. Weasels, white-tailed jack rabbits, snowshoe hares and ptarmigans turn "winter white" in direct relation to periods of snow cover. The hares at Ellesmere Island, 80 degrees N. Latitude, stay white the year round, but at Hudson Bay they take on a summer coat of brown from late June to late August. In the Green Mountains of Vermont they are brown in mid-June at 2,800 feet, in lighter fur at 3,200 feet, and in fur spotted with white at 4,000 feet, where snow lasts until May. Natural selection long ago set these patterns of protective coloring (which can be induced out of season under controlled lighting, irrespective of high or low temperature). Shorter days bring on the molting of brown into white fur or feathers, and then, after a three-month winter period, longer days are responsible for the return to brown.

While animals are molting, they are extremely vulnerable to predators. The fact is dramatized by a recent note in the *Journal of Mammalogy*. While hunting elk in the White River National Forest of Colorado one day in October, a government biologist saw a marten chase a snowshoe hare with some brown pelage still showing out of a stand of dead spruce into a snowy clearing. There the hare took shelter under a lone spruce, but not for long. Its flash of movement had attracted the attention of a red-tailed hawk roosting on a high branch of the same tree, and the bird of prey began to shift downward from branch to branch, listening and looking for another movement. It was a checkmate situation. After several false starts, the hunted animal rushed into the dense woods in a headlong dash to escape its enemies on the ground and on the wing.

Like the first bluebird of spring, the timing of the first snowfall often varies from almanac predictions. Winter really arrives not on December 21st, but with the first cyclonic meeting of northern and southern air masses that binds ice crys-

Winter Wanderer
When prey is scarce in the North, where this bird usually ranges, the snowy owl sometimes drifts hundreds of miles southward, feeding on any rodents it can catch.

A Cloud of Pollution
Ridges of the Coast Range cut through billows of smog blanketing the Los Angeles region. Manufactured by sunlight from tons of contaminants released daily into the atmosphere (mostly automobile exhaust), the smog disintegrates clothing, corrodes steel, and often promotes human respiratory ailments.

tals and water drops into uniquely patterned snowflakes and sends them eddying and swirling down.

Despite the warming influence of ocean currents on the Atlantic Coast below Cape Hatteras and on the Pacific Coast of California, North America is the snowiest of continents. Most of the annual precipitation north of 37 degrees N. Latitude is in the form of snow, ranging from about six inches collected by the western Great Basin in Nevada to sixteen feet or more on the high mountain peaks. Drifts to the leeward of strong snow fences along tundra ridges in the Rockies are literally reservoirs. The fences retard melt-off, ration the rush of little rivulets that supply major rivers on the Atlantic and Pacific sides of the Continental Divide, and spread part of the springtime surfeit of water through the summer.

The snow blanket settles unevenly, drifting highest in valleys and in the lee of ridges and lakes. It is a warm cover—sometimes 35 parts of air to one of ice—into which ruffed grouse, rabbits and mice tunnel.

In January, under the weight of snow and ice the boughs of conifers bend within the reach of hungry deer. Ice scars the trees as it expands and contracts under sharp temperature changes, and mice—deprived of other food—girdle the trees where they gnaw at the bark. The more the wind blows, the greater the risk of freezing, even in a fur coat. The lynx with its great snowshoe feet, and the wolf and the coyote that are lightfooted, all can skim over the drifts; but the wapiti and the deer find themselves restricted to small, trampled "yards." There is no way for them to run from danger without floundering, and therefore many of the weak and the young die now, in the severe cold of the weeks after the solstice.

The coldest part of the earth is Siberia, caught annually in a deep freeze with little snowfall. Similar conditions exist on parts of the Great Plains, where winds sweep the ground bare of snow and the freeze goes down to a depth of six feet, prolonging winter. In the area of heavy snows around the upper Great Lakes, there is a ground freeze of only 30 inches. This means an early thaw, a leisurely "growing season" of long days, an abundance of life.

Where does spring begin? Perhaps, as Edwin Way Teale suggests, in the "watery wilderness of the Everglades," where the flowers of spring and fall bloom together in December. "Somewhere south of Lake Okeechobee," he writes, "spring comes into being, swells, gains momentum. Its arrival becomes more abrupt, more striking, its line of demarcation more evident as it progresses north."

Cosmic events have gone full cycle, marking their effects on many different landscapes and providing plants and animals with an unvarying rhythm of darkness and light, winter and summer, under which a living cell can be wound up like a tiny clock to grow, reproduce, or merely survive for another year.

In the past, the radiation balance of the earth and its climate zones have scarcely varied for thousands of years at a time. Nothing that man might do in his cultural evolution from stone tools to the gasoline engine could change the laws governing the exchange of heat between the atmosphere and outer space. But the Industrial Revolution and the explosion of human population during the last hundred years is beginning to have an effect.

Around urban concentrations of belching smokestacks and automobile exhausts, an eye-smarting, lung-irritating, and sometimes deadly pall of smoke and gases forms during periods of natural inversion in which cool air rises above warmer air. Although particles of grease, oil, and metal oxides in the atmosphere are the most obvious pollutants, invisible gases such as sulphur dioxide cause more death and property damage. When "smog" is at its worst, the ultraviolet, or sterilizing, rays of the sun are screened out, permitting viruses to spread.

How can we cut down on the noxious smog? Nuclear power plants, mostly in the blueprint stage of development, will give off only krypton, a gas that can be stored and released under favorable weather conditions. But coal and oil—our primary fuels of the present—each contain sulphur; when burned, they pollute the air with sulphur dioxide or droplets of sulphuric acid. Some relief is possible, because not all heavy oil has the same sulphur content. High-sulphur Venezuelan oil that powers much of our industry in the Northeast can be diluted with less sulphurous Libyan or Nigerian oils. A difficulty, however, is that some low-sulphur foreign oils cannot be refined economically.

An alternative to oil as an industrial fuel is natural gas, which contains virtually no sulphur. But unfortunately the burning of gas produces nitrogen oxides, setting up another pollution problem. The widespread use of fuel gas instead of oil in Los Angeles has probably added to the photochemical smog arising in that city.

The Angelenos have taken measures to reduce many kinds of pollution that threaten New York, Pittsburgh, Washington, Baltimore, Chicago and other metropolitan centers, but they have had great difficulty in controlling the soupy brown exhaust mixture, largely hydrocarbons and nitrogen oxides, that is brewed on their freeways. Anti-exhaust devices are only beginning to keep pace with the cars, another two million of which will probably be on the roads of California by 1980.

Air pollution is not limited to cities. The corrosive vapors from smelting works kill plants and likestock and affect the health of people in such remote places as the Montana grasslands and the Arizona desert.

Some scientists see the possibility of cataclysmic events resulting from the burning of fossil fuels in our civilization. These fuels represent the carbon in living and dead organisms accumulated over a period of 500 million years, carbon which is now being returned to the atmosphere at a much higher rate than under natural conditions. From 1860 to 1959, the carbon dioxide in the air increased about seven percent. In the next few decades it will probably rise many times higher—

perhaps enough to increase temperatures and drastically shift the world's climate and rainfall belts.

This is what is known as the "greenhouse effect." A rise in temperature of the lower atmosphere due to the presence of increased carbon dioxide tends to increase evaporation. Augmentation of water vapor in the atmosphere in turn means increased absorption of infrared rays and therefore an even greater heating of the atmosphere. If the process is carried too far, icebergs might melt and the ocean level rise as much as 200 feet, drowning the cities with all their smog problems. Or greater precipitation in the North and on mountain peaks could bring about a new ice age.

Since either alternative would be catastrophic, the possibility of prevention is worth considering. Dr. Roger Revelle, one of the world's foremost oceanographers, has suggested that artificial warming of the earth's atmosphere could be countered by artificially reflecting more of the sun's rays from the planetary surface. This might be done by spreading a smoke screen of finely powdered latex or of a light-colored opaque plastic over the tropical oceans. At an estimated cost of $500 million a year, we might change the radiation balance of the earth in the desired direction by one percent—and, as a by-product, weaken the destructive energy of tropical storms.

But it is entirely possible that the plan could bring about unforeseen results that would make the cure worse than the original disease. We have learned in many ways the hazard that lies in man's attempting to manipulate nature.

2
Wave-Swept and Sheltered Shores

Rocky Coast

Sand Beaches

Mixed Rocky
and Sandy Coast

Coral

Like the specks of dust in a sunbeam, or the stars in the firmament, countless billions of minute plankton organisms float in the sunlit layers of the sea. The majority are plants, composed of a single cell or a few cells linked together, which utilize the energy of the sun to produce carbohydrates. They comprise vast fields for the foraging of animal life. Among the most fertile areas for the production of plankton are bays and estuaries, supplied with the necessary minerals by the outpourings of rivers. But the major supply of nutrients for plankton comes from the ocean depths. The pattern of ocean currents and seasonal changes determines when and where the greenest fields will grow.

Off the coast of Southern California, the prevailing northwest wind of spring and summer blows the surface current away from the land and causes a strong upwelling of deeper, mineral-rich waters in which plankton swiftly increase. Also highly productive in spring and summer are the waters of the Labrador Banks, where two great ocean currents—the Labrador Stream and the Gulf Stream—converge. Here in the North Atlantic, winter storms help to mix ocean layers in advance of the growing season. As the hours of daylight lengthen, tiny plants called diatoms multiply and cloud the water in a seasonal flowering, to be consumed by the hordes of plankton animals. Now the dinoflagellates come into their own. These strange plankton organisms, which propel themselves with whiplike tails, are equipped to live as plants or animals—or both. In the fall, one last, smaller burst of diatoms ends the cycle of growth and reproduction.

All through their season of activity, the plankton are carried inshore and swept out to sea again in a rhythm corresponding with the rising and setting of the moon, which governs the ocean tides. The greatest tidal changes occur twice during the lunar month, the 28-day period in which the moon passes elliptically around our planet. Because it is closer to us, the moon exerts a stronger gravitational pull than the sun, whose bulk can only add to or subtract from the lunar effect.

After the full moon, and again after the new moon, the waters of the Atlantic Ocean bulge higher and send stronger surges against the continental shores because the sun and moon are aligned with respect to the earth, and their attractive forces become added together. During these periods, incoming flood tides rise highest, and ebb tides fall lowest. They are now called "spring" tides, from the Anglo-Saxon "springan," meaning "to leap." In its quarter phases, the moon out in space stands at right angles to the direction of the sun, and the counter-pull of the two bodies lessens the movements of the tides—which are then called "neap," a word that originally meant "hardly enough." Thus the fairly regular tides on the Atlantic shore of North America—with each low water following the next at an interval of 12 hours, 25 minutes (or twice each lunar day)—become greater or lesser during the month. The pattern may be changed locally by the shape of a peninsula that intercepts the surging waters, or by the rising of an offshore wind that can push an incoming tide back out to sea.

In the Pacific, which is a much broader and deeper ocean than the Atlantic, the moon pulls the water up in a greater bulge and sends generally higher tides to the shores. But it does so with less regularity. The tidal pulse at Victoria, British Columbia, averages 30 feet, and low water occurs only once a day. Along the rest of the West Coast, tides arrive twice a day, and one high tide may follow another in just nine hours. Months may pass without any spring tides. Only in midsummer and midwinter does the sea touch the highest cliffs and expose the lowest kelp forests, and then the spring tides continue for weeks at a time.

Heavy waves add to the height and depth of the tides on the Pacific shore, creating an "intertidal zone" which is far more extensive than that of the Atlantic shore—with the exception of the Maine coast. There the power of the surf increases seasonally with the fury of winter storms, and tides range very high indeed; at the south end of the Bay of Fundy, the water rises to a height of 50 to 70 feet as it passes through a narrow strait.

The greater the tidal range, the wider is the band of seaweeds and shore organisms that anchor or glue themselves to rocks along the coast. The long, white-sand beaches of New England, the Carolinas, Florida, California and Oregon do not provide any equivalent foothold. Surging over the beach, waves stir up the sand grains and sweep this debris along shore or out to sea, and animals dig down instead of trying to hang on.

Wherever there are tides, shore organisms have developed "clocks" precisely timed for the arrival of food, and these inner mechanisms continue to operate even if the rhythm on which the organism depends for survival is arbitrarily changed. When mussels were moved from Cape Cod to California and exposed to Pacific tides that differed by six and a half hours from the Atlantic tides, the mussels opened to feed at the wrong time for a full week before adjusting to the new tidal rhythm.

Many marine organisms, including the grunion

of the West Coast and the palolo worm of the Caribbean, breed according to the seasons—or simply in response to temperature change—but spawn by the phases of the moon. The grunion makes its way onto the shore in time to ensure the return of its progeny to the sea on an outgoing tide. But no one knows why the worm should respond to the lunar cycle by leaving its dark coral retreat and rising to the ocean surface for spawning. The answer is rooted somewhere in the origins of life in the more extensive seas of earth more than half a billion years ago, when tides were probably greater and moon pull may have been the single most important environmental influence on living creatures. In ages long past, changes in pressure on gonads, turgid with eggs or sperm, triggered spawning, and the inherited "psychic pattern" continues under the weaker tidal regime of today—not just in sea worms and fishes, but also in land mammals. In the females of certain primates, including human beings, the menstrual rhythms are still geared to a cycle of 28 days, though no longer strictly tied to the waxing and waning of the moon.

What happens when the sea, whose pulses bring life to the special world of the shore, is corrupted with the wastes of civilization? Sewage feeds a poisonous dinoflagellate, one of the causes of the "red tide" that paralyzes shellfish and sickens the man who eats them. More dangerous is the waste from nuclear reactors. Even when it reaches the sea in small quantities, the waste matter accumulates in the tissues of those same shellfish, and in other organisms as well, just as phosphorus and other essential elements do—and with unknown implications for the future.

Some 375 million years ago, molten rock pushed up the earth's crust to form the core of the Mount Desert Range of Maine. Gradually the material cooled and solidified into a coarse-grained pink granite, and much of the overlying sedimentary rock washed away. Then great glaciers bore down, overtopping the mountains and quarrying a new landscape. During the Wisconsin Period, the latest and most frigid of the glacial episodes, a 5,000-foot-thick sheet of ice plowed deep valleys through what had been a nearly continuous mountain ridge running east and west. As it rode over the south faces of the mountain peaks, the glacier sheared off rock walls and giant steps. Under the weight of its advance, the plains to the east sank below the sea, leaving above water only the wave-bound cliffs which we know today as Mount Desert Island. Once the ice retreated and its pressure was gone, the earth slowly began to recover and rebound. The change is recorded in tilted cliffs and in tide-gauge measurements which relate land level to ocean level. These measurements show that North Atlantic shores are still being uplifted—in some places having risen as much as 12 inches in the last century.

On the sheer cliffs of Maine's Mount Desert Island, the tidal ranges are among the highest in the world. Here the sea steadily grinds out caves and chasms from the hard granite—especially where the surf is heaviest, as at Thunder Hole and Anemone Cave. But in the few thousand years since the last Ice Age, the waves have done little to alter the Maine coastline, which is not so much eroded as *invaded* by the sea. Where the long, narrow estuaries of the Kennebec, the Sheepscot, the Damariscotta and other rivers run inland for many miles, the rocky, forested ridges that overlook the drowned valleys have been shaped mostly by ancient stresses from within the earth itself and by glacial chiseling and weathering. Now the cold of winter and the heat of summer—as well as rain water and intruding plant roots—are continuing the destruction of peninsulas and outlying islands, slowly breaking the rocks down into sand and pebbles, which the waves carry to little beaches in the recesses of inlets and coves.

On the rocky shore, the sweep of the tides and the pounding waves shape the form and behavior of life. The rhythmic rise and fall of the sea, between the extremes of high and low water, bring oxygen and food twice each day, more or less regularly, to the inhabitants of the mid-tide zone. These are sea creatures which have developed the ability to endure exposure to the air, either by sealing themselves tightly into shells or by taking shelter, as in the case of starfish and anemones,

The Land Between Tides
On a Maine Coast, white barnacles mark the reach of high tide and seaweeds darken the low-tide area. Between these limits live shore organisms of the mid-tide zone.

in moist rock niches. Above them are the white encampments of barnacles, touched by waves and spray only a few times each month. Here the rhythm is catch-as-catch-can. Only when the highest tides roll in do the barnacle's plates open. Then the animal's feathery nets emerge to sweep up the plankton on which it feeds, in a brief period of intense activity. After the tide ebbs, the organism can only close its plates and hold moisture about its body, staying quiet until aroused by the next watery advance. Higher still on the rocks are the periwinkles, which have pushed above the spring-tide line into the splash zone, and can survive a month's separation from the sea.

Far below, in the low-tide zone, uncovered for only a few hours each month, the animal communities resemble those offshore in deeper water. Out there, the competition is keener among a greater variety of species. The strictly shore-dwelling creatures have worked their way up the rim of the continent, out of the reach of hordes of fishes and other swimmers—except during those times when predators are brought in by the tides that also bring plankton, the universal and indispensable gift of the sea.

Nothing better illustrates the primordial ascent of life from the sea than the zonation of three species of periwinkles on the New England coast. The smooth periwinkle, found nestled among wet seaweeds between tidal pulses, can stand no more than a brief exposure to the air, the sun and the wind. Another, the common periwinkle, often lives where it is submerged for only a short time at high tide. Both drop their eggs into the sea on the high spring tides, which will return many of the larvae to their proper levels. But the rough periwinkle, that lives high up in the splash zone, has cut most of its ties with the ocean. It is viviparous, and its larvae settle on the shore without first having had to survive as defenseless eggs drifting among carnivorous plankton-feeders. As adults, they are able to exist largely out of the water because, unlike their relatives below, these snails possess gill cavities well supplied with blood vessels that function almost like a lung to breathe oxygen from the air. Constant immersion would be fatal for them. If they are swept away by waves and deposited at a lower level, they find their way

A Crumbling Coast
Cliffs of hard granite along the North Atlantic shore of our continent are highly resistant to erosion by waves alone. But extreme temperature changes, in summer and winter, may contract and expand the rock, cracking it along internal pressure lines into almost symmetrical blocks (near right). Later these cracks are enlarged and rounded by the action of water (center photograph). Then a colorful crust of lichens can take hold, further splitting the rocks with their roots (far right). Thus cliffs are transformed into boulders, eventually to become pebbles and sand.

back up, drawn by increasing light and dryness.

The rough periwinkle scrapes microscopic plant food from the rocks with a rasping organ, as do limpets and other grazing snails. It pastures among dark fields of blue-green algae. These, the most primitive and the hardiest of marine plants, have emerged all the world over to colonize rocky shores above the barnacle zone.

The farther an animal lives from water, whether the distance is up a cliff or at the inner end of an estuary, the more necessary to survival is its memory of a tidal rhythm. The rough periwinkle's biological clock is set for a peak of feeding activity twice each month, at the spring tides, with periods of desiccation and sluggishness in between. When taken into the laboratory, snails of this species continue to act in accordance with the old rhythm for many months, even though artificial "tides" bring them food more frequently.

The vegetation of a sheltered shore grows in zones somewhat like the climate zones on a mountain, where plant life diminishes toward the frigid summit. Successive zones are banded with different seaweeds, each more resistant to

periods of drying than the one beneath it. Above the level of the high spring tides hang the vivid, yellow-green strands of spiral wrack, delicate and dried by long exposure. Below, in the mid-tide zone, lies the brown forest of rockweeds, with long, trailing fronds. The lower rocks and the walls of tidal pools are thickly overgrown and matted with the red seaweeds, dulse, and Irish moss. In the very lowest pools, just on the verge of the ocean, the huge deep-sea weeds, or kelps, begin to appear.

The pattern of shore life varies from place to place, depending on the force of the surf. On open, wave-lashed shores, the seaweeds only appear far below a broad band of barnacles and mussels. No seaweed can survive the big surf on the Atlantic headlands and off-shore islands. But the barnacle, being small and permanently cemented to its rock, is well-suited to a surf existence. So, too, is the limpet, which crawls about on a broad "foot" with a suction that can resist a pull of 70 pounds per square inch. The conical shell of this snail offers no resistance to waves, which roll over it harmlessly. The shape develops

as the larva grows on the rock, from a spiral to a flattened Mexican hat—showing a direct relationship between form and the forces of the environment. When the snail is at rést, its foot secretes a corrosive acid that carves out a niche in the rock, just the right size for the animal, and to this niche it usually returns and rests after grazing.

When not fishing with its flowery tentacles, the anemone retracts into a round ball and holds on to the rock with a rubbery suction cup so tightly that a wave would have to remove a piece of the shoreline to dislodge the animal. The blue-black mussel spins a number of fine threads that hold it down, much as guy ropes secure a tent; in heavy surf, the streamlined prow of the mussel swings about, taking the shock head-on at its point of least resistance. Among the closely packed ranks of the mussels, spiraled whelks and other carnivorous snails take shelter during heavy storms.

On the shoreward side of the outer reefs, some seaweeds grow, while flat colonies of sticky sea squirts and tough, limy sponges feed by filtering the spray. Their form is austere. The same animals in more sheltered places proliferate in pendulous masses of soft jelly and fingery lacework.

The reproductive rate of encrusting shore organisms, from tiny tentacled hyroids to grapelike clusters of sea squirts, is tremendous. Many generations are produced during each spring and summer. But only a few adults survive the summer months, and even the hardiest of these may be swept away by storm waves and ice during a severe winter. Then several years will elapse before colonies of hundreds and thousands re-appear on the rocks and pilings.

November 1 is the beginning of hibernation for most shore animals, and the first sign of spring within their habitat is the early spawning of a cold-water acorn barnacle in February. Like its summer-spawning relatives, this barnacle matures later and lives longer than most other members of the seashore community; its life expectancy may be three years or more. Barnacles do not leave it to the currents to carry sperm and randomly fertilize their eggs, as some creatures do, and close clustering is a necessity for reproduction. In this and most other species of barnacles, one of the feathery tentacles functions as a penis, stretching several inches to probe the open valves of neighbors on every side. Thus cross-fertilization takes place. Eggs develop in every mantle cavity, and milky clouds of larvae hatch into the sea.

In its first stage, when it is only one hundredth of an inch long, the acorn barnacle resembles a miniature horseshoe crab. It grows quickly on a diet of the tiniest of algae. Six times the animal's shell is outgrown and shed, and finally, at the end of about 60 days, the barnacle emerges as a bivalve with six legs and two antennae that secrete a special glue. It then drifts or walks about on the substratum for hours or days, testing spots here and there before settling. Perhaps these young are drawn to existing colonies of barnacles through chemical attractants given off by the adults. Or perhaps they just stop whenever the glue hardens. No one knows. After "choosing" a spot, the larva goes through a metamorphosis as complicated as a butterfly's; in just twelve hours the head, the feathery appendages, and the plates of the cone are complete. The animal will never move again. No other crustacean elects to be fixed by the top of its head in a castle of lime, and then lie in wait for the rest of its life for the returning tides, as the barnacle does.

One of the dangers of an immobile life is vulnerability to enemies, and barnacles have many of them—including snails, fishes, and even birds. The chief predator on these shores is a large carnivorous snail, the dog whelk, which is equipped with a long proboscis that it uses both as a drill and a feeding organ. Usually, however, the snail does not bother to drill a hole, but merely envelops the barnacle with its fleshy "foot" and applies strategic pressure. Once the barnacle's valves are forced open, entrance into the cone becomes easy. The dog whelk is also armed with a secretion called purpurin—the source of Tyrian purple dyes, long taken from a related whelk in the Mediterranean—that may act as a narcotic to subdue the snail's victim.

By their voracious feeding, a herd of dog whelks can cut a wide swath of destruction across the millions of tiny, sharply pointed cones on the upper shore, greatly altering the chalky appearance of the rock surface. In one area observed, so many barnacles were killed that mussels moved in, darkening the shore with their clusters.

A Shore Dinner

*Cemented permanently to the rock where it has settled,
a barnacle (right) is dependent upon the tide to
bring in food. Opening its plates, the barnacle sifts
the ocean water with feathery appendages for
tiny plankton organisms. Its enemy, the predatory
dog whelk, roams at will along the shore. Forcing open
a barnacle with the powerful suction of its "foot"
(below), the whelk consumes the barnacle's tender meat.*

A Seaside Nursery
In a rock crevice at the mid-tide level hangs a cluster of the yellow egg capsules of the dog whelk. One whelk may produce as many as 245 in a single season, each of the capsules containing about a thousand eggs.

Life in a Tide Pool
Between rockweed holdfasts in the wet, sheltered world of the mid-tide pool, a conical limpet travels over red and yellow algae, and a pink young starfish settles among dark mussels and periwinkle snails.

The dog whelks promptly switched to the new food. In making the switch, they spent many futile days boring holes in empty mussel shells, some even drilling from the inside. Eventually, however, they ate the mussel colony out of house and home. Barnacles repopulated the rocks, and the whelks then returned to their old diet.

The colors of whelks—orange and white, white and purple, or pure white, lavender and brownish-black—reflect what they have been eating. The pigments of mussels can be seen in the shells of the darker whelks (often striped), which contrast sharply with the white, barnacle-fed individuals.

In the rock pools at the mid-tide level, empty barnacle shells often house the larvae of anemones and periwinkles as well as baby barnacles. All around hang the brown draperies of rockweed, the curtains of a shore nursery where whelks wedge their capsule eggs into protected crevices, young horse mussels hang on by silken threads, miniature starfishes cling with tube feet, and tiny crabs skitter in sidelong dashes from frond to frond. This jungle is as full of food as it is of danger. The starfish, with special sense organs on the tips of its rays, has no trouble finding the clusters of young, tender mussels, especially if the shells of some are broken so that the mussels give off a detectable odor. With the same organs, the starfish carefully feels its way to avoid falling into the trap of an anemone, a creature which is nothing but an expansive gut surrounded by stinging tentacles.

Only a small proportion of the little crabs, starfishes and other young produced in this crowded zone ever reach maturity. One medium-sized crab carries as many as 4,900,000 eggs. Few of these will hatch, fewer will escape their predators in early life, and fewer still will survive to spawn. Yet the competition for space is so intense that barnacles settle on other shelled organisms, both mobile and sessile, and even on seaweeds—just as they foul ships' bottoms and pilings. Lower down, mussels settle on oyster beds in numbers so great as to smother the oysters. Deep-water barnacles may then become attached to the mussels, killing them in turn. Most of the echinoderms (starfishes, sea urchins and their relatives) have evolved little pinching organs all over their bodies to discourage such invasions of privacy.

Like the kelps of the dim and shadowy deeps, the seaweed invaders of the low and mid-tide zones are pigmented with brown or red—colors that mask their chlorophyll and absorb most efficiently the warm range of the spectrum (essential to photosynthesis), which diminishes as light becomes weaker under water. Since their leaves are bathed in the nutritious salts of the sea whenever the tides flow, these seaweeds have no use for absorbing roots. But their holdfasts branch out like the roots of a forest tree, anchoring each plant against the possibility of floating away or succumbing to wave shock. Light determines where the anchor develops on the floating "egg" of the rockweed. On the side that is exposed to sunlight a leaf grows, while the shadowed side produces a holdfast. Along the rocky shore these holdfasts may be only inches apart, rising in arches over a carpet of small green algae and the limy secretion of a red seaweed which stains the rock. In the dark, wet shelters are myriads of shore babies, as well as limpets which graze the algae, scraping trails so deep that they swallow particles of rock along with their food. When the tides are out, the fronds of rockweed hang limp, dripping moisture onto their inhabitants and protecting them from the sun.

The mid-tide pool provides a small asylum to which salt-water organisms may retreat. (Higher up, there is the danger of dilution by rain, or of evaporation by the sun, which crystallizes salt around the edges of exposed pools.) Dense growths of algae in the bottom of the pools produce so much oxygen by photosynthesis that streams of bubbles rise to the surface in the daytime, yet after dark there may be an oxygen deficit. When the waves rush in on the crest of the returning tide, there is a complete interchange of brackish and new salt water in the pools, restoring the conditions for filter-feeding—and, in the spring of the year, for recolonizing every level of the shore. In the tidal surge, the bladder wrack floats on vesicles filled with air, and knotted wrack takes wing on its finely-serrated blades. Below these streaming fronds, the holdfasts now teem with predators from the sea, grabbing a quick lunch before the tide ebbs, to leave here and there a ray or a jellyfish stranded in the sun.

Where a Wrong Move Means Death
A starfish, immobilized by the stinging tentacles of a sea anemone,
is engulfed in seconds, disappearing into the stomach cavity through which
the anemone both sucks in its food and expels undigested matter.
Having an insatiable appetite, this creature will try to swallow almost
anything that comes along, sometimes including pieces of shell.

Wave-Sculptured Rocks
*Chimney-like stacks of rock, in the foreground, stand before massive
sea arches in a small cove south of Morro Bay, California. Storm waves of the
Pacific break down the less resistant rocks of the coastline first, often
tunneling out an arch. When the roof of such an arch falls in, a stack is formed.*

Gathering strength over an area twice as large as the Atlantic, the storm waves of the Pacific Ocean are hurled against the West Coast with an impact of up to a ton per square foot of rock. The force of the water continuously carves out caves, arches, and isolated stacks from the soft, volcanic rock of the projecting headlands. The Jesuit Father, Don José Campoi, has suggested that the name California must have been derived from the Spanish word *cala*, a small cove of the sea, and the Latin word *fornix*, an arch. The two words together perfectly describe the panorama of wave erosion from Monterey Bay to the lagoons of Baja California, which were visited by Cortez during the earliest Spanish explorations. This is an emergent coast of sheer cliffs standing on wave-cut platforms, some barely covered by the present tides and others well above the spray.

Disturbances within the earth raised these shores during the late Pleistocene Epoch, when the shores of Maine were sinking under the weight of glaciers. Unlike the Atlantic, the Pacific coast lies within an area regularly convulsed by earthquakes and the sort of slow but persistent mountain-building activity that raised the Himalayas to their present height in a million years. Scattered in a random pattern from about 10 to 70 miles off shore, the Channel Islands are volcanic mountains whose submerged flanks rest on underwater sills. In this area a general subsidence of the shoreline left only eight small mountain tops above the waves, and numerous submarine canyons, varying from about 2,000 to 6,000 feet in depth. Some of the fault lines along which the earth slipped to create these deep rift valleys may be traced from the California shore to points a thousand miles out into the Pacific basin, and many islands formed by volcanic eruptions have come and gone without leaving a trace except for lava and ashes at the bottom of the sea.

The almost vertical cliffs of the offshore rocks and the Channel Islands are occupied in the summer by nesting sea birds. Colonies of cormorants, brown pelicans, and California gulls crowd the tops and squeeze into every available rock niche below. From August to September the skies are filled with northern terns, gulls, scoters, loons, and a few shorebirds whose migrations follow the rocky headlands. At certain points near Monterey, San Diego and Santa Barbara, one can glimpse

lines of sheerwaters traveling by the millions far out over the sea. Here, in quiet coves, are the nurseries of the California sea lion. Five other marine mammals breed or winter on the more distant offshore islands. San Nicholas and Santa Rosa are among the last sanctuaries of the Southern, or Guadelupe, fur seal and the elephant seal, both nearly exterminated in the waters off Mexico and Southern California by 19th Century sealers.

Among the most traveled of all animals are the Northern fur seals, which leave their arctic rookeries early in June and make an ocean circuit through the deep waters about 50 miles off the Pacific coast. This species barely survived the sealers, but has come back from 200,000 in 1911 to a million and a half under controlled hunting. (The harvest of fur that takes place annually in the U.S. Pribilof Islands is limited to a certain number of four or five-year-old males.) Occasionally a blue whale—the largest living mammal, weighing up to 150 tons—wanders up from the antarctic. The arctic gray whale, which weighs about 35 tons, glides past San Diego in an annual migration of about 3,000 animals, widely spaced

in small herds. A century ago, the gray whales passed at the rate of 1,000 a day during January, on their way to breeding lagoons along western Baja, Sonora and Sinaloa, Mexico. But so many were slaughtered by West Coast whalers that, like the right whales of the Atlantic, they practically disappeared. Some were subsequently discovered off Korea, on another migration route followed by whales and seals, and recently sizable herds have come back to North American shores under the protection of the International Whaling Agreement of 1937.

Recently a new threat has developed. Gray whales returning to the arctic in the spring of 1969 found themselves in the midst of the Santa Barbara oil disaster—the worst to hit any coast since the wreck of the tanker *Torrey Canyon* off Cornwall, England, two years ago. Although geologists knew the area was honeycombed with fault lines, some of which leak oil naturally, a series of giant oil-drilling platforms has sprung up in the channel, each with 60 to 100 shafts penetrating the unstable ocean floor. The shafts run about 3500 feet through beds of mixed shale and sand into pressurized oil pools. To avoid blow-outs, the

Bird Colony on a Rock
*Protected by an ocean moat from all enemies on the ground,
double-crested cormorants nest in profusion on a rock island off the
California coast. On the mainland, where foxes and other carnivores are a
threat to their eggs and young, these birds nest in trees. But here
at Morro Bay, pairs crowd together (see below) in safe little territories.*

state and federal governments provided guidelines for the construction of metal casings to depths of about 1200 feet or less, depending on the subsurface conditions. The possibility that these regulations might be inadequate was not even considered. Late in January of this year, one of the Union Oil Company's wells on federally-leased Platform A blew, and a great volume of gas and oil (4200 barrels a day) bubbled up to foul the sea, the harbors and beaches and even the air. In the frantic days that followed, oil drilling was halted twice while all the companies working on the federal leases conferred with the new Secretary of the Interior, Walter J. Hickel, about stricter regulations. For weeks, Union Oil sent soap boats to Platform A. The spraying of detergents, together with choppy seas, helped to break up the ugly dark flow as it diminished. But vagrant slicks, floating

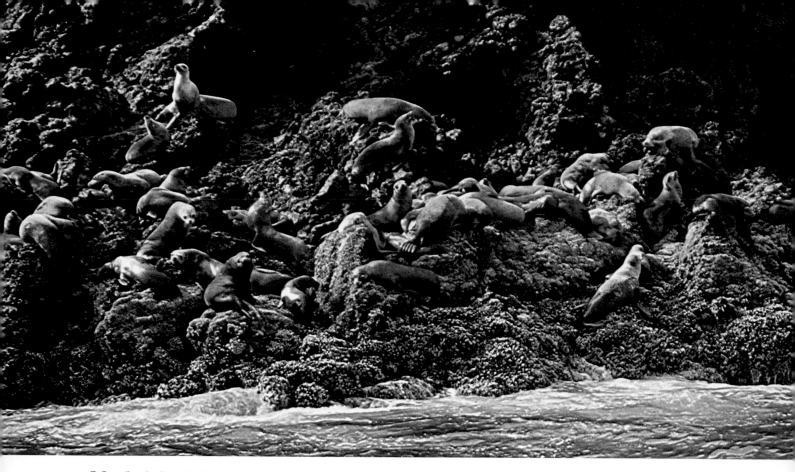

Island of the Seals

Ten miles out in the Pacific, foggy Anacapa Island harbors hundreds of California sea lions in its small, rocky bays. The big bull seals, weighing as much at 1,000 pounds, gather their harems of females in this favored breeding location from May to August, and the seal pups are born a year later.

on the currents all the way up to Point Conception, trapped entire flocks of water birds. Most seriously affected were diving birds, especially Western grebes, which winter in the channel. No one counted the corpses, but oil company crews combed the beaches and picked up a total of 1700 oil-soaked birds that were still alive. These were carefully washed with a mild detergent, polycomplex A-11, and dried under infrared lamps. The only trouble was that the survivors could not be put back in water—with the natural oil gone from their feathers, they sank. It was difficult to feed most of the birds in cages, and at the end of February, all but about 100 were dead. By mid-March, some gray whales were stranding on the Santa Barbara beaches, their mouths full of oil—the heavy, black crude oil that had settled on the channel bottom where the big cetaceans feed on shrimps and mollusks.

In the wake of the disaster have come teams of biologists, searching for other broken strands in the shore communities, which are less obvious than dead birds and whales. The leak, though controlled, still remains a hazard, and, like Pandora's box, cannot be completely sealed without danger of mounting gas pressures and possibly another, bigger blow-out. So another well is necessary to tap the deep pool and—it is hoped—prevent the recurrence of the black tides.

The migratory birds, seals and whales that make a highroad of California waters, moving down from the arctic in the winter and up from the tropics in the summer, are following a cycle of winds and currents. In summer the wind over the Pacific blows out of the north, deflecting a warm, northward-moving coastal current out to sea. To replace this water, a deeper, colder countercurrent wells up to the surface and moves southward along the shore. But in the winter the wind shifts to the south, the upwelling subsides, and the warm current returns, to travel northward at a speed of about half a mile an hour.

From year to year the winds can change the pattern. If the northwesterlies arrive early in the spring, the summer will be cold. The cool sea and air temperatures on the ocean side of the Coast Ranges are met by heat from the Interior Valley, and as a result, heavy fog envelops the headlands each morning—slowly retreating out beyond the breakers by noon. The harbinger of such a season is the *Vellela*, a small plankton animal with a diagonal sail like a piece of cellophane, which strands in large numbers on the shore. If the northwesterly winds fail to develop, the weather stays mild and sunny, with little fog.

The over-all climate of the Pacific coast is mild compared to the Atlantic coast. Sea and air temperatures from Seattle, Washington, to Baja California range, on the average, from 52 degrees F. in winter to 61 degrees F. in summer. In the absence of extreme winter cold and summer heat, the activities and breeding seasons of Pacific animals are less restricted than in the East. But a slight variation in the narrow temperature range to which they are adapted can be catastrophic. This is most critical at the diatom level of the food chain.

Diatoms—microscopic "pill boxes" with transparent walls of silica—are very specifically oriented to cool, temperate waters. If the water warms too much, bacteria increase and destroy the cells. If the water cools too much, the diatoms sink, dead or dormant, to the bottom of the sea. Only under optimum conditions of light and temperature does cell division, the diatom's principal method of reproduction, begin. The top and bottom of the tiny box draw apart, and the nucleus splits, forming two identical boxes, each with the proper number of greenish-brown chloroplasts to help the plants obtain energy from the sun. When actively dividing, one diatom may have a million descendants in 30 days. A tiny shrimp-like copepod, floating along in the same current, may eat 120,000 diatoms in a day. And 60,000 copepods have been found in the belly of a single sardine.

Most of the birds and mammals that gather along the Pacific shore feed on sardines, anchovies, and other small fishes. Suddenly, not long ago, a major question on the West Coast became: "What happened to the sardines?" In the mid-1930's, the annual catch was reckoned in millions of tons, and Monterey's Cannery Row—immortalized in the novel by John Steinbeck—was in its halcyon days. Then the fisheries suffered a sharp setback. Catches diminished irregularly for a decade, until in 1947 the sardines almost disappeared. Nine years of investigation by American, Japanese and Canadian research teams disclosed some significant facts. The northwesterly winds were blowing stronger and the ocean had become colder than in the peak sardine years of 1920–1938. The difference in ocean temperature was enough to delay the hatching of sardine eggs by many hours and thus retard the development of the fry. Besides posing a direct threat to the survival of the young, the lowered temperatures provided conditions for a rapid increase among their predators, the zooplankton, which became 20 times more abundant. Furthermore, strong upwelling and current action may have dispersed the fry to places unfavorable for growth.

The survey revealed that the sardines had shifted their spawning far to the south, off the coast of Baja California, and that even in these waters they were scarce. Then, in 1957, the prevailing winds shifted. Typhoons developed over the South Pacific, ocean temperatures rose, southern species of fish turned up in fishing nets far north of their usual ranges, and there was a slight upsurge in the sardine catch. It seemed apparent that weather changes, causing variations in sea temperature of only two or three degrees, had brought about the near-demise of the sardine industry. The changes affected men no less than whales, seals and sea birds—all occupying positions at the ends of coastal food chains.

At this point in Southern California's history, little is left of natural life along the shore. The caves have been cleaned of their abalones by divers; the tide pools have been poisoned with copper sulfate by fishermen harvesting octopus; and the seals have been killed or driven offshore by hunters. Yet within the gates of certain strictly fenced seaside ranches, nature remains undisturbed. Below Morro Bay, on a single stretch of private coastline, you can walk up close to baby sea lions sleeping in the sun, and can watch double-crested cormorants incubating their eggs on an offshore rock, their colony separated from the mainland by only a few feet of ocean. Along the cliffs, black oystercatchers dive on thick mussel beds between recurring sea spouts.

For perhaps an hour or two the tide withdraws from sea caves lit with the reflected brilliance of sunbeams on coralline algae. Here the low-tide pools are inhabited by green sea anemones second in size only to those of Australia's Great Barrier Reef. In the most protected places they grow up to a foot in diameter, and their powerful tentacles, with a suction of 15.6 pounds per square inch, can capture a small fish. In the dim interior of a cave all the anemones are white; those in the sunlight are bright green. The latter harbor a single-celled algal plant (*Zoochlorella*) in a symbiotic relationship. The algae receive carbon dioxide, a waste product of the anemone's metabolism, and the anemone uses oxygen, a by-product of the algae's photosynthesis, to "burn" food within its tissues. Both plant and animal benefit from their close cooperation; but the algae, completely contained and protected, may have slightly the better of the situation. The anemone must still obtain some of its oxygen from the surging tides.

In these caves there are not likely to be any red abalones. The big, commercially marketed snail disappeared from the intertidal zone long ago, and it hangs on—somewhat precariously—only in the offshore kelp forests, 30 to 50 feet or more under the sea. California periodically bans out-of-state shipments, thus reserving for residents and tourists the delicacy of abalone steak—the snail's meaty "foot." Half a dozen smaller species inhabit California shores, but few can be profitably harvested. The green abalone, heavily exploited during the early days of the industry, has declined. Since 1948, the pink abalone has supplied more than half of the annual catch, and is the principal abalone gathered by skin divers and other sportsmen. Another species, the black abalone, has been saved by its habits. It clings to crevices where heavy surf pours plenty of oxygen-filled water over the foot and adjacent gill chamber of the animal. The gills expel this water through an unusually long row of open portholes in the shell. From having to hold tightly against the surging sea, the black abalone has developed a muscular, tough foot, which requires a great deal of tenderizing before it is edible.

All along the beaches, broken bits of abalone shells add to the usual debris and flotsam an iridescent glitter that is missing from Atlantic shores. Like the lining of an oyster, the rainbow-colored interior of the abalone shell is secreted by the animal, and a pearl may form around some irritating grain of sand or parasite, usually a tiny boring clam. As time goes on, more and more concentric layers of pearly shell are laid down. Fisheries studies indicate that red abalones grow to maturity (about 4 inches in diameter) in six years, and reach legal size for commercial harvesting (7¾ inches) in twelve years. The oldest, which measure about 12 inches across, are among the largest in the world.

The community at the low-tide level is dominated by the California mussel and the goose barnacle, the latter a creature out of mid-Seventeenth Century English bestiaries. Its lanky stalk was said to bud and give birth to geese. Differences in appearance aside, goose barnacles open up and sweep the sea with feathery nets just like their relatives, the cone-shaped acorn barnacles. Among the goose barnacles crawls the common starfish, searching mostly for mussels, but also preying on barnacles, snails and sometimes small crabs.

The closely packed beds afford many niches for smaller animals. In the spaces between and underneath the mussels and their "guy ropes" are trapped particles of decaying matter on which the little flat crab and the worm *Nereis* feed. Other creatures scrape the film of algae that grows on the mussel shells and on the rocks beneath them. As the community grows, there is room for a huge variety of encrusting organisms. One patch of rock ten inches square was found to contain 4,711 animals of 22 different species. Perhaps the closest associate of the mussel is its messmate, the pea crab, which shelters in the bivalve's mantle cavity and combs its gills for scraps of food.

Outer, exposed shores are colonized mainly by the mussel, the goose barnacle, and the common

An Ocean Cave
Ebb tide in a sea cave below Morro Bay, California, uncovers abalones and seaweeds that flourish in the dark interior, protected from wave shock, sun and wind.

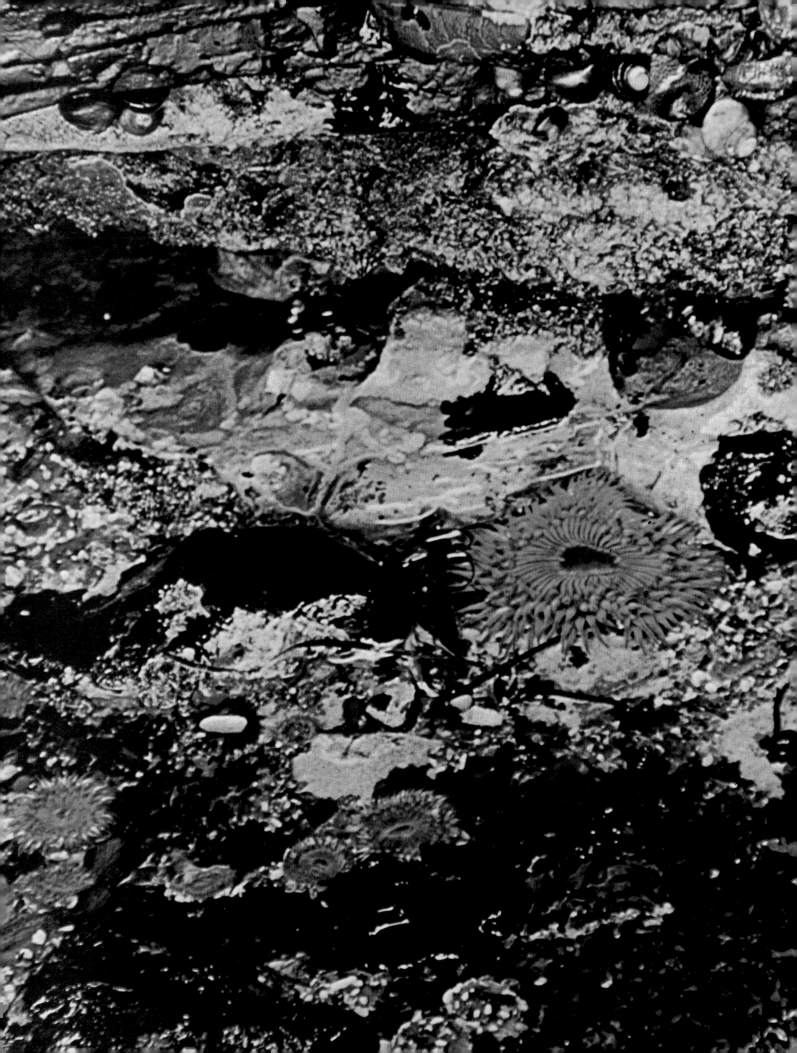

The Scavenger
As the tide ebbs in a sea cave on the California coast, a lined shore crab emerges from beneath a curtain of damp seaweed to search the tidepools for edible scraps.

A Star-Shaped Predator
Feeding among mussels, a common starfish applies strategic pressure to open one of the bivalves, then forces its stomach into the mussel's shell to digest the flesh.

An Emerald Trap
An oral disc surrounded by poisonous tentacles is the green anemone. It waits for food to be brought in with the returning tide, which enters the sea cave with a gurgling rush (opposite page), sweeping over a bright collage of organisms.

starfish, without their animal associates of the protected, low-tide level. Under the surf-swept rocks, tube-worms build extensive honeycombs, picking their building materials right out of suspension in the waves. Bits of broken mollusk shells, echinoid spines, angular quartz grains—the tube-worms glue them all together with mucus in an overlapping spiral pattern that ends in openings only eight to ten millimeters wide. From these openings, plumed heads emerge in unison and comb the generous spray.

Nearly all of California's rocky headlands are awash with the brown blades of a giant kelp, native only to western North America and the cooler shores of the Southern Hemisphere. In his classic *Voyage of the Beagle*, Charles Darwin compared these seaweed forests, anchored in rock just beyond the breaker zone, with the great tropical forests of the world. He predicted a chain reaction of death in the sea if they were destroyed, a prophesy that has proved accurate in the 20th Century. In recent years the growing coastal cities have discharged an ever-increasing amount of sewage—now calculated at half a million gallons a day—into the offshore waters, causing algae blooms and kelp die-offs.

Few plants grow and reproduce as abundantly as kelp. In the submarine stands, the long, hose-like stems shoot up to the surface at the rate of two feet a day, to shade mile on mile of the ocean floor with their lengthy fronds, buoyed on clusters of hollow bulbs. Each plant emits a constant stream of spores (up to seventy trillion annually), insurance that some will drift to a sunny spot along the kelp periphery and grow, expanding the forest. Underneath the canopy the sea swells more gently, and in the dark calmness there is a clustering of shellfish and fishes, an accumulation of marine life that attracts birds, sea otters and seals. No one knew how delicately the strands of the food web were interwoven in the great stands of kelp until entire forest patches shriveled up and were washed away in the late 1940's—much to the consternation of chemical companies, skin divers and fishermen.

As early as 1910, kelp was harvested for potash, an important fertilizer. When that market became

A Neck Like a Goose
The little goose barnacle, shown here several times larger than life, withdraws its telescopic neck until a wave arrives. Then, stretching its neck, it puts out a feathery net at the end to capture microscopic plankton.

unprofitable, the chemists discovered another product for the synthetic age. Algin, the substance that gives the seaweeds their toughness and flexibility, is now used to suspend, stabilize, gel, and emulsify a wide variety of prepared foods and pre-mixed medicines. To some people, the death of the kelp forests posed a threat to the supply of "gel" for their ice cream. To the skin divers, it meant the loss of a playground for spear-fishing in which black bass often grew to the formidable weight of 400 pounds. The commercial fishermen missed a dependable source of market fish. And the abalone hunters no longer could hope to discover those masses of great dark shells hanging onto the sides of submarine canyons so tenaciously that it took a tire iron to remove them.

Almost everyone blamed the chemical companies for over-harvesting the kelp; but the chemists pointed accusingly to sewage outfalls. Many years passed before the mystery could be solved. In the meantime the weather changed, and warmer-than-usual summer waters intensified the recession of kelp. By the late 1950's, the Palos Verdes forest off Los Angeles, which once covered three square miles, was a virtual marine desert. The Point Loma forest off San Diego, originally six square miles, shrank to less than one. Wheeler North, then of the Scripps Institution of Oceanography, began diving for clues, and discovered herds of bottom-hugging sea urchins, as many as one hundred per square yard. Urchins, which are grazers with five tiny microscopic teeth, usually move in "fronts" through kelp, never staying long enough to destroy the forest. But these urchins were sedentary; "subsidized" by sewage particles and scum algae spawned by the sewage, they stayed and gave the kelp no chance to regenerate. What could be done about them?

One of the few predators that can stomach the spiny little animals is the sea otter, the world's rarest and most valuable fur-bearing animal. But hunters along the Pacific shore destroyed so many sea otters during the early years of exploitation that only one sizeable herd —about a hundred individuals—has returned to the coast, south of Carmel. And though protected by law, these potential allies of man are still killed sometimes by abalone hunters who mistakenly regard them as competitors.

An accidental oil spill cleared one urchin-in-fested area. Quicklime treatments also proved effective in getting rid of the urchins, after some trial and error, without harming fishes or abalones. Now the forests are returning, at least temporarily, with the help of an elaborate transplant program. Kelp is towed from flourishing areas to places with a suitable rocky bottom five or ten miles away, or else the precious weed is grown in the laboratory on nylon ropes and then laid down. The success of the project depends on the kelp's quick growth. If not eaten by fishes grazing on the isolated stands and young plantings, new forests may grow up in a year.

They may, that is, if there are no further setbacks. A dying forest of kelp off La Jolla started to revive after a quicklime treatment, only to succumb to a "red tide" of dinoflagellates which clouded the water and deprived the kelp of needed sunlight. An increasing number of atomic power plants along the coast pose a further threat. Their discharges may raise ocean temperatures above the survival point of the giant kelp, which thrives only in cool waters. Scientists, in their search for solutions to the problem, foresee a time when a transplanted tropical kelp from Mexico may replace the native species, just as native oyster beds depleted by over-harvesting have been seeded with Japanese oysters.

Exacerbated by battles between private interests, the many problems along the shore will probably go on generating new problems until all the people concerned combine forces and look at the picture in the large. One hope is the Sea Grant College Bill, signed by President Johnson in 1966, under which marine preserves are being set aside, often in cooperation with industry. Such projects show the way towards consolidation of plans for beaches, marinas, oil drills, nuclear reactors and waste disposal lines, while also providing for clean ocean areas—areas such as the proposed underwater wilderness off the Scripps Institution of Oceanography in La Jolla. Seventeen miles long and half a mile wide, this state preserve would include two unique submarine canyons.

Scripps Canyon and other offshore deeps are suspected of swallowing up Southern California's sand beaches, which have been disappearing at an alarming rate in recent years. Along the coastline runs a littoral current as powerful as a great river, steadily carrying away more sand than arrives from "upstream." Man-made dams in the coastal ranges impound ocean-bound sediments, and a twenty-year drought has severely reduced all runoff. With supplies depleted at the source, massive and repeated sand hauls are necessary to restore the beaches and dunes, which are even then mutilated by beach-buggy tracks and strewn with discarded automobile parts. In only a few areas is random driving across the dunes restricted so as to leave intact some natural barriers against the advance of the sea.

Near San Luis Obispo, onshore winds pile up ribs of sand as pure as the Sahara's, with a swirled central backbone 300 to 400 feet high. Behind them, the gentler, more rounded dunes are anchored by tough beach grass with dense root mats and by masses of the starry, rust-tinged ice plant and pink verbena, in a landscape dotted with small fresh-water ponds and tule marshes. This is Dune Lakes, a private sanctuary on the edge of the Pacific.

Although the ever-shifting dunes of pure sand are relatively lifeless, the more stable, grassy dunes shield a bustling community of insects closely tied to one another as prey, predator or parasite. Tiny aphids suck the juices of the plants. Various ants "milk" the aphids, and the aphid wasp injects into their bodies eggs that hatch into larvae which will eventually consume the aphids and pupate in their empty shells. In the daytime the ants and flies that feed on the nectar of the flowers fall prey to stinging sand wasps, and robber wasps emerge after dark to hunt katydids and crickets.

The surf-swept sand beach below the dunes contains few animals, and these are specialized burrowers. Some, like the mole crab and the razor clam, dig in very rapidly. But the Pismo clam and others with heavy structural reinforcements depend on thick shells to avoid being crushed, and on ribbing to hold them against the strong currents. The Pismo clam not only tolerates such currents; it cannot survive without them—as commercial clammers discovered when they tried to store their catches in lagoons and sheltered bays before shipment. Invariably the shellfish died. Through generations of living in constantly agitated water, the species now requires more oxygen than do clams in quiet bays. It has developed a very fine net of branched papillae across the open-

A Sea of Sand
Miles of sand dunes, rippled by Pacific winds, extend along the coast near San Luis Obispo, California. They serve as a barrier to shore erosion—and provide a rare panorama of unspoiled natural beauty.

ing of its siphon, a screen that keeps out the swirling sand and at the same time lets in water and microscopic particles of food.

Many years ago, the entire broad, intertidal sweep of Pismo Beach, lying just north of Dune Lakes, was a rich clam bed, and teams of horses drew plows through the sand, turning up clams by the wagonload. As a result, diggers must now wade out waist-deep or even deeper in the surf to find any clams at all. The Fish and Game Commission makes an annual census, based on test counts in strips of beach running from the upper limit of the tidal zone down into the zone of extreme low water. The Commission reports indicate that, despite present restrictions which severely limit the season and allowed catch, the once-abundant Pismo clam is in danger of becoming extinct. Like the abalone of the rocky coast, this shellfish grows slowly, requiring four to seven years to reach its present "legal size." It continues to add dark growth rings to its shell until the clam is at least fifteen years old. Except for man, its enemies are few. Starfish stay away from the surf-swept sands in which Pismo clams live, gulls take only the smaller ones that are washed up, and the hard shells of the adults resist all of the boring snails.

During the heat of the daytime, dwellers beneath the sand stay as cool as the organisms under the seaweeds on a rocky shore. Signs of hidden life vary from the tiny breathing holes of nocturnal insects to the "wakes" left by the eye-stalks of mole crabs as they move about at mid-tide. These crabs, equipped with large, feathery antennae, feed only as water runs down the beach. As a wave travels shoreward, hundreds of mole crabs emerge from the sand and allow themselves to be carried along until, at the end of the surge, they hastily dig in and put out "fishing" nets. The entire colony moves up the beach several times with the incoming tide, and the reverse occurs when the tide goes out.

The mating habits of the mole crab are singular. On the end segment of each of the four legs of the male there is a suction pad. Before the female lays eggs, three or four of the males usually find her. Clinging to her carapace by their toes, they swirl through the water, attaching ribbons of sticky sperm to her body. About twelve hours later, when the eggs are laid, she will use the sperm to fertilize them.

Also governed by tides are sandbeach isopods, which stay buried until after the crest, and then emerge and swim about in the swirling waters for

Buried up to the Eyes

A flounder buries all but its eyes in the bottom mud, and waits to surprise its prey—shrimps and small fish. The flounder begins life with eyes on both sides of its head, but soon one eye starts moving to the other side. The fully developed fish spends its life swimming or burrowing in a flat position, with both eyes up.

 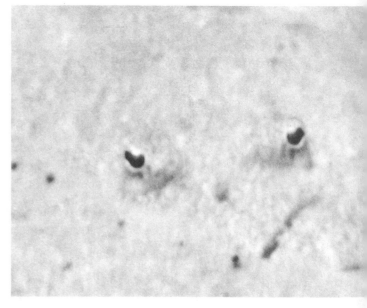

four to six hours. An investigator at the University of California in Los Angeles recently collected some of these tiny crustaceans and put them through all kinds of environmental cycles: different feeding regimes, periods of submergence and exposure, shifts from light to darkness. He even exposed them to cycles of chemical stimuli and oxygen tension. None of the stimuli, however, prompted as much activity in the isopods as swirling water, especially when the water was full of sand.

On the more protected sand and mud flats, the tidal currents bring in quantities of sea weed, which is broken down into a decaying muck that becomes added to the organic wastes from the many plants already growing there. Any movement stirs up this material and brings it within reach of hungry mouths. By siphoning in particles of organic matter and minerals, and then ejecting some, shellfish serve the very important function of recycling essential nutrients and keeping them in the environment for the use of the whole community. A colony of mussels, for example, will circulate every two and a half days as much phosphorus as the water can hold in suspended particles, as well as a considerable quantity of dissolved phosphorus. Therefore a constant amount of this material remains on the surface in any given area of flats, and is not carried out by the tides.

Sulfur, the principal constituent of amino acids—and therefore of proteins—comes largely from the decomposition of the bodies of plants and animals on the bottom layers of the flats. The action of decaying microorganisms releases hydrogen sulfide. Some of this compound is reconverted to a usable sulfate by specialized sulfur bacteria, many of which are chemosynthetic—obtaining their energy from this chemical reaction (the oxidation of sulfur) instead of light. During one part of the sulfur cycle, when iron compounds are formed, stores of phosphorus, trapped in oxygen-poor layers of mud, may be released. Thus, one cycle meshes with another, and a single reaction can change the balance of the whole chemical environment.

In the quiet, settling waters of California's bays and proliferating marinas, bacterial action may be setting off "red tides," poisonous plankton blooms similar to those that frequently foul the Gulf Coast of Florida and the coast of Peru. The offenders—all dinoflagellates with red eye spots—are normally to be found in the calm, blue-green depths of tropical oceans, where they float in widely separated, milky clouds, a feast for fishes and baleen whales. Some, however, range into cooler waters, where they often turn into a scourge.

Though we understand little of the complex chemistry of the sea, the red tides seem to be caused by a sudden increase in certain elements that are ordinarily present only in traces. Heavy rains, such as occur in Florida, may wash these elements into the sea, or the elements may appear as a natural by-product of the growth of organisms. Bacteria and blue-green algae, for example, both produce vitamin B_{12}, a cobalt compound necessary for the nutrition of dinoflagellates. A small amount of this compound triggers a modest and harmless dinoflagellate flowering. But heavy enrichment of the water will cause a bloom so dense that the aggregation of red eye spots colors the sea for miles. It is some mysterious by-product of the bloom, not yet identified, that kills living creatures. One recent outbreak on the West Coast left 10 million fish floating, belly up, in the bays, and even caused the death of a captured pilot whale in an ocean-connected lagoon of the Sea World Oceanarium in San Diego. Three hundred and forty-six people were treated for shellfish poisoning, and twenty-four died. Subsequently a ban was placed on the harvesting of mussels, clams and other bivalves from May to September—the dangerous months of contamination, when the over-fertility of the coastal waters may bring on a red tide.

North of Monterey, the coastline of California is cool, moist and foggy. Here mountains covered with redwoods and other conifers overlook bays, islands and dunes. In the 1850's the first redwoods were cut at Humboldt Bay, and after World War II a new plywood industry ushered in the Douglas fir boom. Across the California countryside hundreds of small mills sprang up, each with its refuse burner and pall of smoke. All along the wild beaches there accumulated piles of waste, carried from the cut-over hillsides on flooding rivers. Now the lumber lands of the Pacific Northwest are in a new era of control by giant corporations which look forward to harvesting a second growth of

Elk on the Beach
A small band of cow elk roams the beach below the Gold Bluffs of
Prairie Creek State Park, California. These are the dark Roosevelt,
or Olympic elk, more often seen in the lush rain forests of the Northwest.

redwood on the bald hills around Eureka and
Arcata. The shores of Humboldt Bay are green
again with new trees, but these will never be
allowed to grow as tall as the thousand-year-old
giants in the memorial groves of the state and
federal parks. Old or new, the forests still extend
down to the sea through canyons of fern and
alders, and at Prairie Creek State Park the Roosevelt elk still range along the marshy ocean fringe.

On the East Coast, the transition from a south
temperate to a north temperate climate is more
dramatic than on the West Coast because of the
influence of Atlantic Ocean currents. The tropical
Gulf Stream follows the continental shelf from
Florida to Cape Cod, and the chill Labrador
Current swings down along the Maine shoreline.
For many thousands of years, the long arm of
Cape Cod has formed a boundary between the
ranges of cold-water and warm-water species of
coastal life—especially in the winter. (In summer, the invisible temperature barriers break
down, and tropical migrants regularly invade the
Gulf of Maine.) Where the currents meet and
intermix, between Cape Cod and Cape Hatteras,
there is a pool of Northern and Southern species,
from which the latter move out to colonize the

North during times of climatic change such as the present era. Among the warm-water animals that have rounded the Cape since Eastern winters became milder in the early 1900's are green crabs, mantis shrimp, and menhaden. Whiting, once only summer visitors to northern New England, now run offshore there the year round. Conversely, some cold-water species—notably the herring—have all but disappeared from the Gulf of Maine.

At Cape Cod the landscape is one of vastly-extending sand beaches, built from the erosion of great glacial moraines. The 40-mile Outer Beach, the Provincetown Hook, and the narrow spits of Nauset Beach and Monomoy are all additions made by the sea at the expense of cliffs that once reached two miles beyond the eastern shore. Over the last half century alone, 160 feet of these highlands at Truro, overlooking the Outer Beach, have been chewed away, mostly by winter surf. From Long Island southward, the longshore currents have piled up the sediments of the Ice Age in numerous barrier beaches, which break the impact of Atlantic storms on the mainland. The wind-driven sand dunes—nowhere higher than at Nag's Head, just north of the National Seashore at Hatteras—are an important second line of defense against erosion of the continent by the sea.

Whatever the waves take away, the currents will put down in another place. And here, as on the West Coast, man's attempts to alter the natural pattern of things have been self-defeating. Jetties built to protect the mouth of Cape May Inlet, for example, trap the tides sweeping out of Delaware Bay and create a whirlpool effect that eats into the coast to the south. In the last ninety years some fifty blocks of real estate have been consumed by the sea, and Cape May Point continues to disappear at the rate of about ten feet a year. In scores of places from Long Island to the Carolinas, real estate developers have speeded coastal erosion by cutting down or leveling sand dunes to build shorefront houses.

From South Carolina to Georgia, the barrier beaches give way to the Sea Islands, a line of sand bars, marshy mud flats and partly-drowned hilltops. For thousands of years the sea has periodically advanced and retreated over the coastal plain in this region, leaving a stairstep pattern of giant terraces extending from the inland hills to the outer realm of the tides. It is from these highlands, past and present, that the rivers and currents have carried mineral sands to Florida's long coast, laying down hard-packed plains of quartz crystals. Toward the subtropical tip of the Florida Peninsula the crystals are mingled more and more with fragments of highly alkaline shell and coral, which limit the abundance of microscopic life in the sand.

Among the numerous animals inhabiting the seashores from southern New England to Florida are fiddler crabs, which herd by the thousands on the quiet flats facing the marshes. Usually two species live side by side, one preferring sand and the other mud. Both the brilliantly-pigmented sand fiddlers and the dull-gray mud fiddlers become darker by day and paler by night, in a rhythm completely different from the "moon clock" that brings them out to feed at the ebb tides every twelve hours and fifty minutes.

When the tide goes out on a sandy beach in the early morning, the snowy-white strand is quickly painted with a palette of intense red, yellow and orange as the fiddlers emerge from their burrows and scurry to the margins of the shore. The male is equipped with a huge claw, the "fiddle," used for display or defense, and a small claw, the "bow," with which he feeds and polishes his "fiddle." The female, which has two small pincers, picks up food with both. It is the male that appears to be "fiddling" while he grooms, and when signaling his readiness to mate. Look closely at the crowd on the beach, and you will notice that some of the males are left-handed and others right-handed. At each succeeding molt, a small claw replaces the big one, and a new fiddle grows on the other side.

The fiddler crabs mill about like a school of fish in the sea. As their claws pick up microsocopic plants and animals from the water's edge, a steady crackling sound fills the air. If they happen to find a dead fish stranded on the beach, there is much jousting among the males; each threatens the others with his fiddle in an effort to capture and keep a favorable position on the carcass, while he pulls off tiny bits of meat with the other claw. When the feeding ends, the crabs return to their burrows, which are dug at a point just above the mid-tide level. The slanting hallways and underground chambers never flood, even when covered at high tide, because the crabs line their homes with wet sand and plug the entrances after retiring. Digging and repairing of the burrows entails rolling up little balls of sand and carrying them to the surface, and sometimes this results in a comic interplay between neighbors. One of the crabs rolls a ball of wet sand over to another hole and sidles back to his burrow, whereupon his neighbor returns the unwanted ball. On a Florida beach one day we watched two fiddlers roll sand balls back and forth half a dozen times, until their "play" was interrupted by the flight of a black-crowned night heron overhead. Instantly the whole community of crabs disappeared down their holes, leaving the beach deserted.

Crabs and crab-eating herons traditionally share living space in the salt marshes, lagoons and tidal flats. These places are increasingly threatened with dredging and filling—especially in the Northeast, where 45,000 acres of marshland were destroyed in the period between 1955 and 1964 alone. Dredging brings parking lots, houses, airports and people; and the disturbance, as well as the actual loss of habitat, has caused many birds to move inland. Although black-crowned night herons still come to Castle Neck, Massachusetts, the great heronries that were there thirty years ago are gone, and it is unusual now to find a single nesting pair.

Beginning in July, the outer beaches of Massachusetts fill up with migrant shorebirds—the sandpipers, plovers and sanderlings that nest around the Arctic Circle during the short polar summer, and afterwards follow the Atlantic flyway to Florida, or sometimes as far south as the tip of South America. At high tide, great flocks of these birds huddle together on the dry, grassy rim of the dunes, awaiting the ebb. When the water has receded far enough, and barely covers the tiny insects, crustaceans and clams at the low tide line, the birds descend and scatter along the foaming fringe. Each has a unique style of hunting. The very small, semi-palmated sandpiper runs about with its head down, dabbing at random. The sanderling walks and probes ahead very deliberately, often leaving long furrows in the wet sand like a miniature plow. More than any other visitor to the shore, this bird enjoys chasing and dodging the surf; the stormier the day, the more food it gleans from the advancing and retreating waves.

The greater yellowlegs is a large, aggressive wader, always on the alert to warn the other birds of danger with loud, insistent cries. Because of this habit, the yellowlegs was often the first to be killed by pot hunters, who entrenched themselves along these beaches every fall before the turn of

The Fighting Fiddler
With a huge claw like a boxing glove, the male fiddler crab guards its burrow in the sand. The crab uses its smaller claw, the "bow," to incessantly polish the "fiddle," keeping this formidable appendage ready to signal and attract a female fiddler crab—or perhaps to challenge another male.

Dancing outside its burrow, a male fiddler crab signals in hope of attracting a mate.

The Territorial Imperative

In a pattern of behavior peculiar to this shore creature, a male fiddler crab stands on tiptoe on a Florida beach and waves its big claw, or "fiddle," in a ritual dance, hoping to lure a female fiddler into its burrow. As often happens among other species, before a female can be attracted another male arrives without invitation, and the two crabs fight. The fiddle is now a major weapon, as each antagonist tries to crush the other's big claw. Straining with fiddles interlocked, the crabs push and pull one another across the sand. The battle ends when the first fighter slips into its burrow and blocks the entrance with its fiddle, thus holding possession of the "territory."

Another male fiddler arrives to challenge the first crab for possession of the hole.

The crabs battle until one establishes its territorial rights by occupying the burrow.

Scavengers on the Wing
*Laughing gulls in winter plumage wheel overhead
on their daily tour of the beach at Flamingo,
in Everglades National Park. No scrap of food
escapes their keen sight and voracious appetites.
If not sufficiently fed on cast-off garbage,
they sometimes eat the seeds of the cabbage palm.*

Competition on the Ground
*Screaming loudly, a young herring gull defends
the remains of a fish from other members of the
flock, which are always close by and waiting
for any chance that might offer to steal a meal.*

the century and shot down dozens of shorebirds at a time. By 1880, the numbers of migratory species had dropped sharply, and nesting species, always fewer, were nearly wiped out. After the Migratory Bird Treaty between the United States and Canada ended mass shooting in 1918, most of the Northern birds came back, though not in the old abundance. (One species, the Eskimo curlew, may be extinct.) Today Plum Island off Newburyport, the Outer Beach of Cape Cod, Monomoy, Pea Island, and Cape Hatteras are all important links in a chain of national seashores and waterfowl refuges, protected against the new threat of shore development.

Among the rarest of shorebirds today are the piping plover, Wilson's plover, and the American oystercatcher, which nest on Eastern beaches south of the Canadian border. A number of other birds in the same family, including the spotted sandpiper and the killdeer, nest around marshes and ponds all across the continent. These inland species have expanded their range into such reclaimed desert areas as the San Joaquin Valley of California, and fatten on the grasshoppers, cutworms and other insect foods of the newly-cultivated farmlands.

Most adaptable of all are the gulls. Unabashed scavengers, they have not only wandered inland but have also increased along the shore, thriving on the refuse of trawlers and fishing docks. Every sewage outlet and garbage dump of our coastal cities usually has its gull attendants, especially in winter. Most gulls nest on small coastal islands, guarded by treacherous shoals and currents, or on remote sand spits unfit for summer homes. Yet even these places were decimated during a period of "egging" and plume-hunting in the 19th Century. The present large colonies along the New England and Middle Atlantic coasts have built up only recently.

On the protected islands, gulls and terns seem to go through cycles, alternating with each other in dominance for periods that may last fifteen to thirty years. Now the gulls are in the ascendency. At dune-covered Muskeget, northwest of Nantucket, there are some 20,000 pairs of herring gulls (probably the largest colony of this common "sea gull" of the Northeast), about 15,000 pairs of laughing gulls, and only 1,000 to 2,000 pairs of common and roseate terns.

An Eye for Hunting

Equipped with owlish eyes, a stubby fish that is appropriately named the short bigeye stalks the coral reef at night, in search of anything that moves among its labyrinthine branches.

Where gulls scream over the eastern keys, at the tip of Florida, the Gulf Stream runs a hundred miles wide and a mile deep, a warm blue torrent with the power of several hundred Mississippi Rivers. This river in the sea originates partly in the North Equatorial Current of the Northern Hemisphere and partly in the South Equatorial Current of the Southern Hemisphere. The latter flows in channels between Caribbean Islands, and curls through the Gulf of Mexico, before finally joining its sister current in the Atlantic Ocean.

It is this southern tributary, funneling out between Florida and Cuba, that has created the underwater wilderness of the coral reefs a few miles seaward of the Florida Keys. The long chain of islands itself is a reminder of coral buildups during ages past. The eastern keys were built by coral animals a few tens of thousands of years ago, when the sea stood perhaps 100 feet higher than today and covered all of southern Florida. In the shallow waters off the sloping southeastern edge of the present peninsula, the corals grew and flourished. Then, during the last glaciation, the sea level dropped and the reef was exposed and killed. Over a long stretch of geologic time, the islands were submerged and then exposed again, so that they now stand above water. Lower portions of the ancient coral are the passes between keys. To the west, the group of keys known as the Pine Islands were formed by sediments that drifted away from the reef, carried by winds and currents. With the changing sea levels, this debris

A Shielded Periscope
*Bumpers of tough shell serve to protect the periscopic eyes of a
spiny lobster, as it makes its way around the coral reef.*

became compacted into a fine-textured limestone, which—because of its resemblance to fish roe—is sometimes known as "oölite."

Coral reefs exist only on coasts or mid-ocean atolls where the water temperature is not likely to fall below 70 degrees F. The corals can thrive (and build up in walls thousands of feet thick) only if conditions are favorable for the secretion of calcium. A tiny animal called a polyp does all the masonry, secreting the material which forms the reef. The live corals grow to a depth of about ten fathoms—as deep as the dinoflagellates that live within their soft tissues can go and still capture sunlight. In the daytime, the photosynthetic activities of these symbiotic partners may provide all the oxygen the corals need. At night, the polyps extend their tentacles to capture the zooplankton that passes over the reef. The current here is strong enough to bring oxygen and food, but gentle enough not to discourage thousands of roomers, large and small, in the grottoes and crevices of the coral structures—which may be branched or boulderlike, flatly encrusting or cup-shaped.

After dark, all of the reef life that is buried and hidden by day ventures forth. Crabs and sea snails of varied shapes crawl out of the gorgonian fans, and spiny lobsters leave their dens under the sheltering bulk of brain coral and star coral. Along the base of the reef wall appear forests of long-spined black sea urchins. Their slender, hollow spines can deliver hornetlike stings. Any sud-

Camouflage in the Reef

To avoid discovery by predators, the spider crab (top) folds up its slender legs and crouches in a crevice. The decorator crab (left) plants bits of seaweed on its shell and hides beneath this "garden." The powerful stone crab (bottom) depends for protection on a shell that looks like a rock and is almost as difficult to break.

The Cleaners

With jaws like pincers, specialized cleaner fishes nibble away encrusting organisms on the bodies of larger fish. Big predators recognize the cleaners and allow them to pick at scales, gills and teeth without being molested. Patterns of bright stripes are worn by both the young angelfish (right) and the adult butterfly fish (far right).

den change in the intensity of light, such as a predator's shadow passing overhead, activates a network of visual receptors scattered widely over the urchin's globular body, and its spines all swivel immediately towards the intruder. The black sea urchin's most persistent enemy is the starfish, which can endure being stuck full of spines and still keep on attacking until its victim is naked and defenseless.

At this depth, if a storm stirs up the chalky white sediments of the bottom, visibility may drop to zero—especially at night. But the big eyes of nocturnal fishes and the periscopic eyes of crustaceans make use of whatever moonlight penetrates the water, and even the weak glow from luminescent fields of plankton. (One reef inhabitant, a hermit crab of the Indian Ocean, carries luminescent sea anemones on its borrowed shell, and thus never travels without a light.) Among the shadows, animals recognize the shapes and movements that mean food or danger.

Nearly all have some place to hide, for the coral is extensively tunneled. Sulfur sponges have dissolved away the calcareous rock. Boring mollusks have riddled it with their workings, and worms with sharp biting jaws have eaten into it. When the tide goes out and exposes the top of the reef, you can hear the claws of tiny shrimp snapping in alarm within the fortifications. The shrimp may come and go, or they may never leave. The circulation of food and essential gases through the canals in the living, breathing coral makes a completely parasitic life possible. The young shrimp, hatched from eggs adhering to their mothers, drift out into the sea and live for a time among the plankton, as do the larval forms of spiny lobsters. Then they return to the coral, or perhaps take up residence on the shallow reef flats in a loggerhead sponge, a large sieve whose intake canals provide the same kind of shelter for lodgers of many kinds—shrimp, amphipods, worms and isopods.

Among the builders of the reef are tube worms and mollusks of the snail tribe, with long spiral casings, whose survival depends on quick retraction. And here, as on other rocky shores, the art of camouflage is well developed. Snails and slender-bodied spider crabs often completely mask themselves with algae and encrusting organisms. The crab instinctively plants this camouflage material, tending it like a gardener.

No creature is really safe, however. Brittle stars with long flexible limbs creep into the sponges and into the little eroded caverns of coral, and eat the minute animals sheltering within. In turn, a small green alga may penetrate the tissues of the brittle star and dissolve its calcareous plates so that the arms break apart, or a parasitic copepod may

destroy its gonads, making it sterile. Armored fishes with strong teeth chew up the coral, together with the mollusks that have disguised themselves inside. And certain angelfish have a way of sidling up and snapping feathery worms right out of their homes.

But the world that the diver sees in the daytime is somnolent and dreamlike. Every flowery coral head has shrunk into its protective cup. Shellfish and urchins seem to become part of the stony reef. Nocturnal fishes bask in the sun or hide in cracks, while others, more active by day, twinkle along the coral façade in a confusing array of colors.

As a gathering place for fishes, the coral reef rivals the kelp forest, partly because so much of the primary plant food here is being produced by very small organisms. The energy value of one gram of algae may be equal to that of many grams of forest leaves. As many as 100 species of fish have been counted around a Pacific atoll, nibbling at the plant-filled reef, grazing its surface, or preying on other fishes.

In a Florida reef, dozens of moray eels of different species and color patterns hide in every coral outcropping, waiting to pounce with needle-sharp teeth on any unwary fish, or on their favorite prey, the octopus. Resembling the sea monsters in medieval woodcuts, these eels are snakelike and sometimes very big (up to ten feet long); they swim about during the midnight hours with crest and dorsal fin just above the water. Outside the

The Fierce Moray
Among the fiercest predators in the world of the reef are these black-edged moray eels with needle-sharp teeth, that lurk in coral caves or hunt along the face of the reef, smelling out their prey.

The Timid Octopus
Often preyed upon by moray eels is the octopus, which cannot escape by hiding in a coral nook.
When attacked, it emits a cloud of ink, setting up a smokescreen to cover its getaway.

reef, sharks and barracudas patrol the channels between the keys.

All of these dangerous predators have symbiotic relationships with small "cleaner" fishes, which nibble the surfaces of their scales, scrape their gills, and sometimes even pick their teeth with impunity. Cleaner fishes are to be found in many parts of the sea, but nowhere in such numbers as in the tropics, where the large fishes are in constant danger of becoming encrusted with harmful parasites, fungi and bacteria. The cleaner fishes get rid of these afflictions.

Usually the cleaner has a definite place of business; that is, it stays near one point on the reef.

It gives a special sales performance—a kind of courtship dance—when a client approaches for cleaning. Without any doubt, the bright poster colors and patterns of many cleaners, including butterfly fishes and neon gobies, help to advertise their profession. The large predators apparently recognize these patterns and come regularly for grooming, and the cleaners seldom get eaten by their customers.

In the coral jungle, however, not all poster-colored fishes are cleaners, and not all cleaners are poster-colored. Evolution seems to have placed no limits on the development of designs here, or on the reasons for the designs. Some, like the

"eye" on a fish's tail, are perhaps meant to deceive; others are probably uniforms of no significance to anyone but another fish of the same species. Konrad Lorenz has noted that the most aggressive of the reef fishes flash the brightest patterns. And all the brightly patterned fishes have occupational specialties. The point-jawed butterfly fish get their food parasitically from anemones and other stinging animals, having learned to steal entrapped prey without getting their own noses stung. Others, completely immune to stings, eat the anemones. Still others stay in the anemone's tentacles for protection.

Each specialist guards its domain vigorously against intrusion by individuals of its own kind. This is especially true of the young, which are often more brightly colored and more aggressive toward each other than are the adults. The males and females of the French angelfish, for example, lose their bold yellow stripes at breeding age. If they did not "lower the flag" of aggression in this manner, Lorenz suggests, they might never be able to mate.

Among more peaceable schooling fishes, such as the grunts and snappers, body colors tend to intensify with age. These fishes and the big tarpons, pompanos, and "lookdowns" that prey on them are seasonal visitors to the reef. They follow the warm currents of the southern seas into the Gulf of Mexico and the Straits of Florida in great processions during the winter and spring.

For centuries, the shoals of turtle grass bordering the passes between the keys have been the feeding grounds of sea turtles, especially the loggerhead and the green turtle. Here they pasture. But at breeding season all of the turtles must swim to distant beaches where the females lumber out of the ocean and, under the cover of darkness, lay their eggs in the sand. This dependence on land has been nearly fatal to the edible green turtles—huge animals, averaging 300 pounds—which have practically vanished from former nesting grounds on the sand keys of the Tortugas and the beaches of Cape Sable in Florida. In all

Looking Down
Knife-slim and fast in the water, with eyes positioned for seeing its prey below, the lookdown fish scans the Atlantic reefs and bottoms from Cape Cod to Brazil.

of their western Caribbean range, only one great turtle bogue remains—at Tortuguero, on the Costa Rican coast. Here every year, from July to September, large herds of females strand on the shore and plod inland, looking for the right spot to dig a hole. Each turtle deposits about 100 eggs. After covering them with sand, she returns to the sea. In about sixty days the eggs hatch, also in the dark of night, and the turtlets, which have never seen water and cannot see it from their birthplace above the high tide line, instinctively head down the beaches toward the Caribbean.

No one knows where the turtlets go during the first year of their lives—not even Archie Carr of the University of Florida, who has been tagging green turtles for about ten years, and who directs a turtle hatchery in Tortuguero sponsored by the Caribbean Conservation Corporation. He believes that they may drift with the currents and live off floating rafts of the sargasso, or gulf weed, which is a complete community of larval creatures, sprouting weeds, and encrusting organisms. By the time they begin to turn up in turtle grass pastures from northern South America to Cedar Key, off the west coast of Florida, the turtles are of "dinner plate to wash-tub size." Some may have migrated from Tortuguero, 1300 miles away.

The adult turtles (they breed when about six years old) have been coming back to the beaches of Tortuguero in greater numbers since the prohibition of "turtle-turning" and egg hunting on the entire Caribbean shore of Costa Rica. Meanwhile, Dr. Carr is raising baby green turtles and releasing them here and on other shores, including Cape Sable. He hopes that they will grow to maturity imprinted by the smell, taste or feel of the place where they entered the sea, and will be instinctively drawn back at breeding time, just as the salmon returns to its hatching place.

In spite of the recent interest in its survival, the green turtle faces an uncertain future. Since the days of the Spanish Main, the demand for its eggs and meat has continuously increased, and turtle soup grows ever more popular. On the turtle grass pastures at Cedar Key and elsewhere in Florida, turtle netting and harpooning are regulated by local laws. But the animal's wandering ways in the open sea make it difficult to protect without international agreements. (In one recorded migration, green turtles journeyed from the Brazilian coast to nesting grounds on Ascension Island in the mid-Atlantic, 1200 miles away.) As the green turtle heads into possible extinction, so may its relative, the loggerhead, whose quiet nesting beaches, from the Gulf Coast of Texas to the Sea Islands of Georgia and South Carolina, are being taken over steadily by man.

The most recent threat to the sea, and to all of its teeming life, is radiation. After atomic devices were fired over coral islands in the Pacific, the radioactive fallout dispersed widely, traveling around the world several times before reaching the ground. Isotopes which had been fused with relatively insoluble particles of iron, silica and dust settled, and are still settling, to enter the bodies of plants and animals through food chains. According to a survey made in the Marshall Islands vicinity, greater amounts of these radioactive isotopes collect in marine plants and animals than in land species. Today, weapons testing in the atmosphere has been virtually halted, but the exploitation of atomic energy for peaceful uses is a continuing threat. Though high-level wastes from nuclear plants, stored in tanks, never reach the land or the water, low-level wastes do— at a current rate of 9 billion gallons a year, with 2 million curies of radioactivity. In terms of gallons or curies, this is not a serious hazard. However, when atomic power becomes a major source of industrial energy in the next few decades, the quantities of wastes—not to mention the "hot water fallout" from the cooling systems of the reactors—will become a major problem along our shores.

Of all the man-made radioactive materials, strontium-90 poses the greatest threat. It is a common component of atomic fallout, and also of atomic wastes throughout the world, and loses its radioactivity slowly, over many years. Strontium behaves chemically like calcium, an abundant and biologically important element. It follows a similar natural cycle through living plants and animals, and is concentrated in the bone marrow of vertebrates. The tendency of organisms to "hoard" radioactive nutrients multiplies the danger of harmful effects. For example, insignificant amounts of radioactive phosphorus released into the Columbia River by reactors on shore were found to be highly concentrated in the eggs of wild geese that obtained their food from the river; the amount of radiophosphorus in a gram of egg yolk was several thousand times the amount in a gram of the river water.

Ten years from now, nuclear reactors in southern California might have a shattering effect on the upwelling of plankton, on the birds, the seals, and—not least of all—the fishing industry. Desalinization of sea water in the Florida Keys, now under way, may be a danger to the reef beyond— or it may be a boon. Who knows whether cooked plankton is tastier to sea life than uncooked plankton? The future of the sea remains a story of change, and one of challenge.

3
The Watery Lowlands

Coastal Plain

Marshlands   Everglades

*O*ver long ages, the sea has ebbed and flowed over the eastern Piedmont, changing the shoreline of North America. In the eons between 11,000 and 25,000 years ago, the continental shelf was exposed and the Northern woolly mammoth ranged in large herds to its very edge. Commercial fishermen and geological survey teams have found mammoth teeth as far as 80 miles offshore in water 270 feet deep. When mammoths, mastodons, horses and llamas grazed the ancient plains of Florida, and saber-toothed tigers and wolves stalked the grassland waterholes, the land must have resembled the Serengeti region of East Africa as it was in the late 1880's. Even during the Ice Ages, Florida weather was warm. The glaciers never reached the peninsula, and the true woolly mammoths never ranged south of Virginia. (But Imperial and Columbian mammoths lived in Florida, and their fossils are among the more common of those unearthed in the region today.)

Members of animal groups that had evolved in isolation in South America when that continent was an island—capybaras, porcupines, ground sloths and armadillos—made their way into Florida after the Central American land bridge was formed in the Pleistocene Epoch. And from the opposite direction mastodons, llamas, horses, wolves and other North American species entered the Southern Hemisphere, where they either prospered or became extinct. The discovery in 1916 of human remains with those of extinct Pleistocene big-game animals near the town of Vero in Florida has stirred controversy ever since about whether Vero Man and his contemporaries caused the disappearance of much of the fauna that had come into the region from two continents. Perhaps, like Folsom Man in the Southwest, Vero Man disputed the ground with the great beasts of the Ice Ages and won, though all such conclusions are speculative unless we find some of his arrows embedded in mastodon bones. Perhaps this was the beginning of the end of the North American game plains and the prologue to the recent disappearance of a long list of birds and animals from the Florida scene—among them the great auk, beaver, flat-tailed muskrat, bog lemming, bison, spectacled bear and jaguar.

During the interglacial stages and in our own time, the level of the sea has risen, covering the sand bars and the river deltas. Today much of the 30-million-year-old Florida plateau, the newest addition to the continent, lies submerged. Freshwater springs bubble up as far as three miles out in the Gulf of Mexico, and veins of artesian waters run thousands of feet deep. The low plains of the interior, scarcely 20 feet above sea level, are cut by countless streams and pitted by some 30,000 lakes—varying in size from small ponds to Lake Okeechobee, one of the largest natural bodies of fresh water in the United States. From cypress swamps and pinelands, the sweet waters run through prairies to the salt marsh and mangrove borders of the sea.

To the west, the coastal prairie of Texas stretches inland for 30 to 50 miles, with luxuriant stands of oak, pecan, elm and cottonwood along the larger rivers. The "big thicket" country, near Beaumont, contains unusually varied forest flora, including pinelands on the dry, sandy ridges, and islands of hardwoods with undergrowth so thick as to be nearly impenetrable. Neighboring Louisiana and Mississippi lie almost entirely on the coastal plain. From elevations of about 500 feet, the rolling uplands drop off to a strip of lush coastal marshes, 10 to 40 miles wide, through which the Mississippi, named the "Big River" by the Ojibway Indians, carries the waters of the continent down an immense alluvial flood plain to the sea. In the drier parts of the interior, pine flatwoods intermingle with oak forests, but in the moist bottomlands grow cypress and other trees that can withstand seasonal flooding. Every spring the Mississippi floods the Atchafalaya Basin in Louisiana, backing water down the many bayous and channels that form natural levees for shallow, marshy areas during the dry season in late summer and fall.

Fingers of the southern Appalachians reach down toward the coastal plain of Alabama and the Florida panhandle, which is primarily pine country. The entire marshy plain from Georgia to Virginia is marked by many shallow depressions that may be the beds of extinct lakes. A few of these "bays," named for the evergreen bay trees that grow in them, are fed by lake-bottom springs. And in southeastern Georgia, two rivers rise from Okefenokee, "the land of the trembling earth." The average level of the swamp, covering about 500,000 acres, is no more than 130 feet above the ocean, of which it was once a part. From the low ridge that divides the eastern from the western section, waters flow off to the Gulf of Mexico by way of the Suwanee River, and to the Atlantic by way of the St. Mary's River.

For thousands of years these lands yielded up by the sea were shaped by wet and dry seasons, by hurricanes that inundated mangroves and marshes with salt water, and by lightning storms that set the brittle grasses and tinder-dry pinelands on fire. These forces are at least as old as the grass and the mangroves, and the natural landscape is a bewildering patchwork of tree

islands in seas of grass, of green mangroves and ghost mangrove forests.

One of the last places to dry up and burn is the cypress swamp—a natural refuge for alligators, cottonmouth moccasins and herons. But even the swamp yields up its wildlife in the droughts that occur about every ten to twenty years. As the water level falls, fishes, amphibians and reptiles perish, and birds fly elsewhere. Peat fires, touched off by lightning, smolder and smoke for weeks.

At the northern tip of the Big Cypress Swamp, in Collier County, Florida, lies a watery wilderness whose 6,080 acres contain the last big growth of virgin bald cypress—the oldest trees in eastern North America. Some of the trees were more than 200 years old when Columbus landed in the West Indies, and are still alive. The bald cypress, a deciduous conifer, loses its foliage in winter and renews its bright green leaves each spring. The tall spires of cypress may reach as high as 150 feet into the sky, and the trunks of these trees have a girth of about 25 feet.

The interior of the swamp is gloomy, mosquito-ridden, and elaborately draped with gray streamers of Spanish moss, but there are occasional openings in the vegetation where giant old trees have fallen. An air plant belonging to the pineapple family, the "moss" lives on the trees but does not harm them. Its hairlike roots, attached to the bark, absorb the water and minerals that drip down the trunk after a downpour. Other members of the same family, garlanding the trunks and branches, look more like true pineapple plants. Their thick, pointed leaves are often furled at the base, forming cups that hold water during the dry season. Insects congregate in these tiny reservoirs, and green tree frogs and snakes climb up the trees to find the water and quench their thirst.

Other lacery that envelops the trees may be unobtrusive (a leafless orchid with tiny "jingle bell" blossoms) or all-embracing (the thick coils of the strangler fig). Often starting at the top of a cypress from a bird-borne seed, the strangler fig sends down long aerials that finally become rooted in the soil. This vine can choke and kill the smaller trees. Giant ferns top the logs, the stumps, and the strange "knees" that develop on swamp cypress—either to buttress the trees or, in

some as-yet-unknown manner, to help the roots "breathe" under water. (Bald cypress on dry sand never grows these appendages, which are also rare in the small pond cypress, a variety found on the outskirts of the swamp.)

Surrounded by red bay, swamp maple, custard apple and other picturesque trees of the understory, the sunlit ponds in the openings are solid-green carpets of water lettuce and duck weed. If you look carefully, you may see the greenery move. Suddenly a snout will rise above the surface and a gaping reptilian mouth appears. This is not necessarily a threat; the alligator and the American crocodile (which lives in the southern Everglades) regulate their body temperature by gaping or by basking in various positions in and out of water. Here the alligators have grown accustomed to people, and often loll unconcerned on logs or on tussocks of vegetation. Over most of their unprotected range, however, alligators are hard to find. Since the 1800's they have been hunted for their hides, and recently poachers in air boats have taken most of the older ones—12 feet and longer—from the vast sloughs of southern Florida. Defying regulations, the hunters speed across the Water Control District above the National Park boundary during the night. When they find alligators, they flash spotlights in their eyes and shoot. The hide on the belly brings six dollars a food untanned, and as the species becomes more and more rare, even refuges and zoos are being raided.

It was not until the alligator was endangered by such activities that anyone realized its importance to the outlying saw-grass areas. Ponds dug by alligators over hundreds of years provide watering holes where fishes survive drought and other animals come to drink. A female alligator maintains one of these 'gator holes by uprooting grass around its edge and constructing a nest for the incubation of her eggs. Unlike most reptiles, which leave the fate of their young to chance, she lurks in the hole, waiting for the eggs in the warm, rotting vegetation to hatch. In about sixty-three days the young begin to break out, calling as they struggle from the shell, and the mother promptly digs them out of the grass and mud. The tiny alligators stay in the excavated hollow for a few days until the yolk sac that nourishes them is completely absorbed. Then they are free

Reservoir in a Tree

Among the tightly furled leaves of a swamp bromeliad slides a rough green snake. The plant, which grows in a tree, has no roots. But at its base it stores as much as four and a half quarts of rain water, providing a pool in which climbing animals and birds can drink and aquatic insects can live and lay their eggs.

to roam the pond. For at least one season—and sometimes longer—the mother protects them from large wading birds, raccoons and bobcats, and from other adult alligators. Once grown to a size of three or four feet, they are independent and have few enemies to fear—except man.

At the turn of the century, the hammering of the ivory-billed woodpecker could be heard throughout the deep swamps of the Southeast. The lumberman's destruction of the big trees, which produced a specialized food supply for the larvae of wood-boring insects, coincided with the sudden disappearance of our biggest woodpecker. The number of these birds dropped sharply after lumbering reached its peak between 1885 and 1900. In 1915, they were seen in about 12 places; by 1930, in only five. Since World War II, they have rarely been seen at all. But late in 1967 three pairs were sighted by ornithologist John Dennis in the

Big Thicket of eastern Texas. He reported that the ivory bill had, rather belatedly, switched to feeding on the insects in pineslashings. This news awakened hope that the species might be adapting to the changed environment.

Harry Goodwin, chief of the Office of Endangered Species of the Bureau of Sport Fisheries and Wildlife, asked timberland owners to rotate their cuttings of pines, thus ensuring a supply of branches on the commercial pinelands. He also asked them to leave pockets of cypress in swamps where they were uneconomical to cut anyway, and to leave some old trees in other areas for nesting. The chances are that the lumber industry will cooperate—as it has in the South since President Franklin D. Roosevelt established Okefenokee National Wildlife Refuge in 1937. The big Georgia swamp had given up 40 per cent of its timber by that time, and more trees had been slashed for turpentine. The ivory bill was gone and so was the

A White Hunter

Its snowy plumage mirrored in clear water, an American egret takes to the air with a fish in its bill. This large wading bird stalks its prey along the shallow margins of pools and streams.

The Snake Bird

An expert swimmer and diver, the anhinga does its hunting under water. When it surfaces with a fish, only the bird's head and long neck are visible—giving the anhinga much the appearance of a snake.

cougar, an animal which still exists in Florida. But the scars are healing. And the bears, raccoons, otters, alligators and other wildlife now have as custodians the descendants of the men who once pursued them.

The disappearing ivory bill somehow symbolizes the lonely stand of cypress in southern Florida, a fragment of its original size, surrounded by the sea of wet grass and the pinelands. In 1954, the Lee-Tidewater Cypress Company was prepared to cut the last trees here, but the National Audubon Society obtained a reprieve. Urgent appeals to the Society's membership produced enough money to purchase nearly all of the old cypress forest from the lumber company, which donated an additional 640 acres. Other lands, forming a buffer zone of pond cypress and pine, were leased at a nominal fee from the Collier Company. This entire area comprises Corkscrew Swamp Sanctuary, a unique natural history museum.

Visitors now explore the steaming interior of the swamp on a boardwalk instead of "bogging it out" waist-deep in the sawgrass, arrowhead and water lettuce. Most, however, come in winter—the dry season—when the trees stand stark and leafless. So they miss the mosquitoes, which impart the true feeling of a tropical jungle, and most of the orchid blossoms. But from December to April they can see thousands of pairs of wood ibises, the only storks on the continent, in an ancestral colony that has been preserved. And the anhingas, American and snowy egrets, great and little blue herons, Louisiana herons and little green herons from the surrounding nest areas often feed here.

During the Victorian Era, when *Godey's Lady's Book* was showing sumptuous plumed hats, sometimes topped with a stuffed bird or two, the American and snowy egrets were prime targets of the feather merchants. Before the feather fad started, about 1875, these birds had been subjected to little but "sport" shooting. Then suddenly they were recognized as the most accessible and the most lucrative of the millinery species. The abun-

A Stalking Alligator
With legs outstretched under the surface, an alligator floats and watches for prey with only its eyes and nose above water. When 'gator holes dry up and the reptiles are crowded together, they may eat one another.

dant heronries of the Gulf Coast and Florida were invaded by hunters who shot the adult birds and left the young to starve. Just as the feather hunters and eggers of New England had taken advantage of the colonial nesting habits of gulls and terns on coastal islands, the feather hunters of the South decimated the colonies of egrets. Only during the nesting season did the birds carry the graceful nuptial plumes—the "aigrettes"—which, at the peak of demand, were worth their weight in gold.

The devastation of bird life is suggested in a passage from A. C. Bent's *Life Histories of North American Birds:* "In 1903 the price offered to plume hunters was $32 an ounce . . . at an auction sale in London, 1,608 packages were sold, each weighing about 30 ounces, a total of 48,240 ounces. As it requires about four birds to make an ounce of plumes, these sales meant 192,960 herons killed at their nests, and from two to three times that number of young or eggs destroyed." The fact that such slaughter took the old and the young indiscriminately and threatened the very survival of the colonies had an unexpected and historic result. It provided the rather disorganized and ineffective "protectionists" of the time with their first strong public argument against the indiscriminate killing of American wildlife. Around the nucleus of a few crusaders grew the American Ornithologists' Union, the various Audubon Societies, and other conservation groups. Gradually the movement gathered strength, and laws were passed restricting feather traffic here and the importation of birds from South America and the Orient. Though laws alone might not have deterred the millinery makers, public opinion—and a change in fashions—did. Many women who had once worn the controversial plumed hats now became ardent bird-watchers By 1914 the cause was won, and the heronries survived to flourish again.

A number of industrialists took up preservation as a hobby. On Avery Island, a salt dome off the Louisiana coast, the late Edward Avery McIlhenny created a pond and heronry in the midst of his elaborately landscaped estate. At first there were only seven young snowy egrets in a large wire enclosure. These grew to maturity; some selected mates, built nests, and raised their own young. Then, at the beginning of the next migratory sea-

son, McIlhenny pulled down the wire, and all the birds flew off to South America, where the snowy egrets of Louisiana normally spend their winters. He hoped that they would return to nest in the spring. So they did—and many other herons moved into the protected area from surrounding marshes. Eventually it became necessary to enlarge the pond to 35 acres and to build platforms of bamboo to support the nests. Carloads of twigs were gathered from distant points and heaped on specially constructed benches from which they could be gathered by the herons. Thus encouraged, the colony soon doubled in size.

Over the years, "Bird City" on Avery Island has attracted thousands of bird-watchers who, from late March to July, follow a footbridge to the tower overlooking the strange scene. Here, where there was no colony at all before 1893, some 30,000 wading birds, mostly snowy egrets and Louisiana herons, are now constantly arriving and departing at the nesting platforms.

In Florida, heronries are beginning to increase in the ponds and sinkhole lakes from the Big Cypress country north to the big springs. To see them, however, it is usually necessary to travel by boat and then slog through muck and marsh. The birds prefer the most isolated shrubby islands, surrounded like castles by watery moats.

Our favorite black willow stand, deep in the backwaters of the Weeki Wachee River, is an integrated housing project of five acres in which many kinds of herons, egrets, anhingas, boat-tailed grackles and smaller birds nest side by side. Only the most recent immigrants, the cattle egrets, keep to themselves—maintaining a small separate community of about twenty pairs. They leave the heronry each day and fly 20 to 30 miles to find their associates, the cattle of the open fields.

Long ago in Asia and Africa these small tropical egrets learned to dig out the ticks from the backs of water buffalo with their sharp beaks, and to pick up insects in the wake of the grazing animals. The domestication of cattle brought an increase in

Little Blues
Fuzzy and awkward at this age, two-week-old little blue herons huddle together in the nest. It will be two years before these white-colored young become dark blue.

the cattle egret population, and eventually the birds began to invade other continents where conditions favored their survival. As recently as 1910 they arrived in South America, perhaps blown there from Africa; by 1952 more wanderers were seen on the east coast of Florida. Having introduced themselves, the cattle egrets are continuing their ancient trade, and apparently intend to stay in the New World.

On our last trip to the heronry, we arrived before sunrise, just as the white cloud of parent birds departed for the day's hunting, leaving the scraggly young to care for themselves—the newborn still nest-bound and others, a few weeks old, clambering around in the branches. Among the most precocious were the long-necked anhingas, or "snake birds." Just ahead of us, a trio of clownish anhinga youngsters flailed about and jabbed at each other with long, pointed beaks. Having no wing feathers as yet, they had difficulty keeping their balance and frequently slipped while trying to hold on to the branches with their webbed feet. Finally, one of them fell some seven feet into the murky water below. For a young heron this might have been a fatal accident, but anhingas are divers, at home in water at first contact. The bird submerged and swam under water, then surfaced and headed for a low, jutting root. Using its beak for leverage, the baby pulled itself out of the pool

and, after a 20-minute climb, managed to reach its nestmates.

In the dark waters, alligators and cottonmouth moccasins wait for mishaps like this that can provide them with an instant meal. Other predators try to penetrate the fortress itself. While we watched, a yellow rat snake which had crossed half a mile of open water climbed up and attacked a nestful of half-grown little blue herons. The reptile struck at the birds, but they dodged and counterattacked with their long, sharp beaks. Fortunately, their enemy's coiled position in the branches was so precarious that they were able to unsettle him and drive him away. Had the young birds not been so big and aggressive, the snake might have succeeded in swallowing some of them.

Most of the wading birds of the heronry eat fish, and the adults predigest some of their catch before giving it to the very young. As the fish is regurgitated, the little bird probes down the throat of its parent to feed. When the nestlings grow larger, whole fish are left for them, and they occupy themselves with squabbling over the food.

The abundance of fish in the heronry—and indeed the whole pattern of life here—depends on the flow from the big spring and subsidiary springs that feed the Weeki Wachee River, lazily winding its eight-mile path to the Gulf of Mexico. Re-charged each year by about sixty inches of

Swamp Babies
*Crowded together in a Florida heronry, young birds of
different species are differently dressed, and all
have distinctive cries and movements that help the
parents find their own offspring among thousands of nests
in the willows. The Louisiana herons (far left)
and anhingas (center) are fed fish from nearby ponds
and sloughs, but the cattle egrets (this page) grow up on
a diet of insects brought from fields many miles away.*

rainfall, underground waters rise from a cavern 50 feet deep at the main spring, pouring out 103 million gallons per day and a tremendous load of dissolved calcium, sulfate, and sodium bicarbonate. The white limestone formations on the shallow bottom of the pool surrounding the boil are directly exposed to the sun and covered with the ribbonlike leaves of eel grass, which seeds and then spreads by sending runners through the fine sand. The delicate female flowers of the eel grass are borne singly on the ends of long stems, flexible enough to sway with the current; the short-stemmed male flowers develop in clusters near the base of the plant. These detach themselves when mature and rise to the surface, where they float. They congregate in great numbers around the females, to which their sticky pollen clings.

Over the entire underwater pasture, algae adhere to the abundant blades of eel grass and entrap diatoms; recent studies have shown as many as 19,000 diatoms to the square inch. And among the blades detritus—particles of decaying organic matter—settles in thick layers. Numerous aquatic insects, snails, herbivorous fish and turtles graze these rich beds of eel grass. A smaller group of fish and invertebrates feed on the grazers. At the very top of the aquatic food pyramid swim bass, garfish and older alligators, the elite of the big carnivores.

Next to the coral reef, the clear, shallow-water communities of Florida rank highest among the habitats of the region in the production of energy. Water birds take a great share of this energy, harvesting the surplus, while alligators, snakes, fish crows and blackbirds live on the young of the water birds in the heronries.

Where the water runs clear, the life of the spring is in balance. A few miles downstream, however, the river banks have been developed and channeled for housing and boat docks. Suddenly, an observer becomes aware of a brown silt—the result of dredging—that clogs and chokes the stream. The silt stops most of the sunlight from reaching underwater algae and hampers the process of photosynthesis. This causes a shortage of food for small fishes and cuts down the oxygen in the water. Without sufficient oxygen, the aerobic bacteria present can no longer digest the organic pollutants—both natural wastes and the drainage from houses. In quiet bends of the stream, oil slicks from passing boats eddy on the surface, adding to the pollution.

The anadramous fishes, those like salmon and shad that spend most of their adult lives in salt water and come to fresh water to spawn, are particularly disturbed by polluted water. They must swim constantly with their mouths open or lie gaping in a current to get enough oxygen. When the dissolved oxygen in the water drops from the normal 15 parts per million in a clear stream to below 4 parts per million, they die. Spawning fishes are particular about the condition of the bottom; if sludge and tin cans have replaced smooth stones, they will abandon entire river systems. Many of our Northeastern rivers are full of industrial pollutants and are too low in oxygen, too mucky, and too poor in practically everything to support any fish except bullheads, shiners and suckers, hardy bottom feeders that "feel" their food rather than "see" it.

The northern part of Florida has been fortunate because of the tremendous storage capacity of its deep limestone aquifers. Though the water table may rise and fall with the fluctuating rainfall, there is always water not far below the surface, and none of the streams has ever been known to dry up. The dependability of the flow protects the streams against drought and stagnation, which have killed shad in the Northeast, and against the threat of floods that seasonally choke up other rivers—especially where strip mining or lumbering has destroyed the vegetation of the watershed.

More water runs below the ground in Florida than above. The vent of an underground river becomes the head of a surface stream, often erupting from the interior of the earth as one large boil and many smaller ones. These flows are nearly all artesian; that is, they are under pressure, corked up under blankets of fairly impermeable rock, like the liquid in a soda bottle. The organic acids and acid-forming gases in the water are the result of high rainfall, an abundance of decaying vegetation, and a low rate of evaporation. Along the flow paths through the porous rocks of the aquifers, large caverns, fissures and cavities form. When the caverns are near the surface, the roofs may fall in, and the resulting sinkholes admit rainwater that helps to recharge the system. This is the way all limestone caves develop, and nowhere are they more open to study than in Florida.

The caves that one finds under the dry slope of a mountain in the American West have long ceased to be "active." The rivers that carved them out have disappeared, leaving open passageways overhung with strange "solution features" created by water dripping through seams and cracks, and floor deposits built up into sculptures resembling people and animals. These beautiful caves are often lifeless, both geologically and biologically. But Florida's caves are still being formed. Many in the northern section lie above the water table, and these dark, wet, air-filled places support a variety of underground creatures that biologists sometimes call troglodytes (from the Greek *dytus*, meaning "to creep into" and *trogle*, "a hole.")

True cave-dwellers never leave the cave, but all have certain vital ties with the outside world. To explore these relationships, we left the pinelands on a humid summer's day that echoed with the thunder of a passing rainstorm and crawled through a six-foot opening in the ground. We entered cautiously, for the pygmy rattlesnake considers the threshold of a cave an excellent spot to catch rodents. Its tiny rattle is like the buzz of an insect—scarcely audible—and its bite, though not lethal, can be extremely painful.

An almost vertical drop, over damp rocks, led to a room that was nearly 50 feet long and just as high. From where we stood, the cave opening now looked like a pinpoint of light. The temperature had dropped noticeably and the air almost dripped with moisture. We descended into another chamber, about 300 feet below the surface, and all traces of light and sound vanished, except for the gurgle of a cool, underground stream. The cave floor dropped suddenly into a tunnel, through which deep water flowed. There was nothing to do but jump in and swim about 30 feet to the next and largest cavern, 300 feet long and 40 feet high. As we emerged, soaked and shivering, a high-pitched, throbbing noise assaulted our ears—the cries of thousands upon thousands of big-eared and pipistrel bats. Hanging from the tall dome one beneath another, in some places they were stacked a foot thick. The floor on which we stood was heavily carpeted with their droppings.

Since no green plants grow in the darkness of the cave, this guano and the bits of uneaten insects brought in by bats are important sources of energy for cave animals. Other nutrients arrive by means of the underground stream, and winds carry in seeds, fungi and bacteria. The food supply may be meager in comparison to that above ground, but there are compensating factors: the temperature of most caves does not vary more than two degrees in a year, and there is little need for activity. In fact, many of the colorless or pale troglodytes in the deeper caverns are sluggish, and some have lost their sight—depending for survival on a very acute sense of touch. Our flashlights picked out a cave cricket and a beetle, both with antennae almost twice the size of the same appendages on their relatives of the surface. These insects and the cave roaches forage in the guano, which grows a crop of fungi and bacteria. Among predators of the insects are blind, colorless cave spiders and crayfish.

Perhaps the weirdest inhabitants of underground streams are the ghostly pink cave-salamanders, which occasionally appear in the outflow of deep wells and surface springs. Only a few of them have been discovered—in the Ozarks of Georgia, and the Edwards Plateau region of West Texas. Some had no eyes at all, and others had only dark eye spots buried under the skin. Although practically nothing is known about their life cycles, it now appears that larvae of the various isolated species hatch out in surface streams near the mouths of caves. They are dark-colored, have perfectly normal eyes, and eat surface insects and worms, like other little salamanders. When the time for metamorphosis comes, the budding troglodytes migrate down into the dark caverns, where they lose both pigment and eyesight. The eyelids partially close or seal, and the eyes degenerate to tiny specks under the skin or else disappear entirely. If the cave waters lack the iodine required by the salamanders for transformation into the adult stage, they may never outgrow their external gills and tail fins. These strange cases of adults in juvenile array parallel those of aberrant salamanders found in certain Western streams.

Some day a salamander may be tagged and traced on its underground trips through caves and crevices between rock strata. If this happens, we shall probably find that its range is very limited. Each cave—or system of caves—usually has its own species of troglodytes, which have become different from those in other underground "islands" through long isolation.

As long as man can remember, there has been seasonal flooding and drought on the Mississippi River. Swollen with snow melt from the north and with spring rains, the river each year overflows to cover adjacent lowlands. Then, towards the end of summer, the river falls, and most of the sloughs and creeks fed by the spring floods dry up. During the 20th Century, however, the

Swallowing a Mouse

Only about a foot long, the pygmy rattlesnake hunts the tiniest of mice, frogs and lizards in the Southeastern pinelands. After killing a mouse, this snake begins the long process of swallowing and digesting it.

nature of the sloughs has changed. On the upper river, tow boats and barges now cruise a 9-foot-deep channel that is maintained, summer and winter, by a long string of locks and dams. These barriers have the effect of stabilizing the water in the bottomlands. Instead of wooded islands and dry marshes in summer, there are now extensive wet marshes filled with aquatic plants.

Eleven of the 26 dams constructed between 1935 and 1939 impound water for our longest wildlife refuge. The Upper Mississippi Wildlife and Fish Refuge follows the river banks from the Chippewa River junction in Minnesota to Rock Island, Illinois, a distance of 284 miles, and encompasses 195,000 acres in four states. Here two government agencies, the Army Corps of Engineers and the Bureau of Sport Fisheries and Wildlife have cooperated to allow boats, water skiers and wildlife to co-exist. The engineers have purchased additional wooded lands to replace those that were permanently flooded. The result has been the preservation of varied habitats for waterfowl, muskrats, mink and beaver. The beaver, re-introduced in 1927, increased so rapidly that trapping was restored, and 13,000 have since been taken from the marshes. Wood ducks, which depend on acorns for food and tree holes for nesting, have been saved, and migratory ducks have benefited. Puddlers feed in the shallow pools, and divers, equally numerous, frequent the deeper water just above the dams.

The lower delta, with its interwoven fields, marshes, groves of live oak, and cypress swamps, is the wintering ground for two thirds of the waterfowl that journey down the Mississippi flyway from northeast and north-central North America. Ducks prefer the low, permanently flooded rice fields and the leveed-off refuges—including 250,000 acres of state-owned land in Louisiana—that have an abundance of wild millet and other annual grasses. Sills installed six inches below marsh level in the Marsh Island and Rockefeller Wildlife Refuges prevent the complete drainage of ponds during periods of low tide, yet do not

Winter Refuge
Regular autumn migrants from inland lakes much farther north, white pelicans descend upon a marsh that stays green all winter in the Mississippi River delta.

obstruct the interchange between fresh and salt water during high tide. In all of the refuges except Pass-a-Loutre, the brackish and salt marshes are burned in the fall to remove the dense wire grass. Blue and snow geese feed heavily on three-cornered grass that grows on the burned-out marsh. Oil rigs share this watery landscape; and the canals and roads built by the oil companies have in the past created difficulties, such as the unwanted drainage of marshes, increased tidal flow, and rapid changes in water levels. Cooperation has been achieved now, and there are special requirements in new leases stipulating that the oil operators use existing levees for their roads.

The main distributary of the Mississippi in the delta region is the Atchafalaya River, which flows 140 miles to the Gulf of Mexico. Formed during very recent geologic times, the river has grown phenomenally since the developments in flood control and navigation. It now carries thirty per cent of the total annual flow of the Mississippi down its channel. Here drifting sediments have contributed to fresh water "ponding," to create rich feeding grounds for waterfowl, deer, muskrats, and imported South American nutrias. During annual periods of high water, practically all of the 422,400-acre basin is flooded. The 20th-Century credo is that nothing should be left uncontrolled. So the Army Engineers have drawn up plans to channel most of the Mississippi overflow and thereby cut off much of a vast wilderness from flooding and sedimentation. Clearly, nature and progress must work out another compromise on the "Big River."

The man-made improvements on the Mississippi have somewhat compensated for wholesale drainage of the nesting grounds of ducks, mostly in the Dakotas and Minnesota. Since World War II, reclamation of U.S. farmlands has eliminated half of the potholes that once covered 115,000 square miles and supported at least 15 million mallards, pintails, and other species of ducks. Today the efforts of farmers to restore the ponds are being subsidized, but the duck population below the Canadian border has nevertheless dropped to less than 5 million.

The survival of the continent's tallest bird, the whooping crane, rests precariously on the fact that its remnant population nests far to the north in the Wood Buffalo National Park of Canada, a wilderness inhospitable to man, and winters on the remote salt water flats of Aransas, Texas. Even before the whoopers felt the pressure of expanding human populations, they were solitary by habit, each family requiring four to five hundred acres for its wintering ground. The total number of birds on the continent reached a record low of 14 in 1938, and each November since then the arrival of migrants in the Gulf Coast refuge has been awaited with some hope and a great deal of concern. Over the decades, there has been an overall gain in population, especially in the past four years. In November of 1967, the Audubon warden at Aransas counted a high of 39 adults and nine young. More important than the total figure is the gain in the number of young that are now successfully making the 2,500-mile journey from Canada to Texas. Mortality is greatest among the year-old birds which are going south on their own for the first time. These are extremely susceptible to the many dangers of migration—including uncertain weather and food supply, and the guns of waterfowl hunters. If the present rate of increase is maintained, however, the species may soon be out of danger.

As a form of survival insurance for whooping cranes, U.S. and Canadian biologists have established a propagation center at Laurel, Maryland. Their plan is to steal some eggs each year, hatch them in captivity, and raise a breeding flock that could eventually be released into the wild. So far, "eggnapping" expeditions in 1966 and 1967 have resulted in a captive flock of eleven birds. The only trouble is that no one knows whether whooping cranes hatched in captivity will breed. Young hatched in New Orleans and San Antonio zoos are the offspring of adults caught in the wild, and all breeding attempts with other species of cranes hatched in incubators have failed.

The most recent threat to animal life in the Gulf Coast refuges has been the chlorinated insecticide, endrin. It was in the bayous of the lower Mississippi that a delayed reaction to endrin precipitated the deaths of an estimated three and a half million fish in 1960, a quarter of a million more in each of the two succeeding years, and a staggering 5,175,000 (including salt-water menhaden in the estuaries) in the winter of 1963–64.

The concentration of the poison was remarkably low—only 0.054 to 0.134 parts per *billion* of water. How could the mass extermination occur? The victims, mostly catfish, had picked up contaminated silt particles while feeding on mud bottoms. They had concentrated amounts as high as six parts per million in their fatty tissues during the summer; then, as water temperatures dropped in winter and the fish were feeding less, they absorbed these fatty reserves into their bloodstreams —and with them lethal amounts of endrin. In the estuaries, the larvae of shrimp were demonstrating a low tolerance to endrin, concentrating extraordinary amounts of it from mere traces in the water. Creatures other than fish were affected. The bayous and backwaters of Louisiana lost countless numbers of turtles, crabs, alligators and otters, along with cormorants and other waterfowl. To this day the brown pelican, the official state bird, has not returned to breed along the coast.

Concerned about the health of the people of the bayous who subsisted on catfish, about the commercial sale of possibly contaminated fish and shrimp, and about the pollution of drinking water in New Orleans, the United States Public Health Service set up monitoring stations. It traced one third of the endrin in the Mississippi to a plant producing the chemical in Memphis, Tennessee, 450 miles upstream. The company was enjoined to clean up a dump which Senator Abraham Ribicoff of Connecticut called a "primitive and dangerous nuisance," and there have been no more winter fish kills of catastrophic proportions. As the heavy use of chemicals on fields of cotton and sugar cane continued in the delta, however, runoff into the streams and pools created pockets of prolific, insecticide-resistant mosquito fish, golden shiners, bluegills, green sunfish and cricket frogs. These were the beginnings of super-contaminated food chains. Because of the clear and present threat to wildlife—and to man himself—all spraying of chlorinated insecticides has now been halted in Louisiana.

With the growing awareness of such dangers, the government has begun taking remedial action of nationwide scope. In 1964, the Interior Department virtually banned the use of DDT, endrin and

A Smooth Splash-Down
More graceful than its bulk suggests, a brown pelican brakes for a landing on the water where it lives by diving for fish. Surviving almost unchanged for 40 million years, this species has recently been decimated by insecticides.

other chlorinated hydrocarbons in all of its refuges and parks. The Dingell-Neuberger Bill, passed by the Eighty-Ninth Congress, increased federal funds for studying the effects of chemicals on wildlife from a former ceiling of $2,565,000 a year to $3.2 million in 1966 and $5 million in each of the last two years.

The great mobility of water birds exposes them to contaminated foods thousands of miles from their nesting grounds, and they may carry the chemicals in their tissues for many weeks after returning to a "clean" diet on the northern tundra. DDT has been found in ducks shot in the North-west Territories at least 500 miles north of any spraying operations, and in their eggs and young. The probable source is insecticide-coated rice in the fields of southern California's Imperial Valley and Louisiana's Gulf Coast.

Although the ducks may survive, the bird of prey that eats them receives a greater concentration of DDT, and the effect on its reproductive mechanism can be crippling to future generations. In a symposium at the University of Rochester in June of 1968, new data were presented that help to explain how this happens. Chlorinated insecticides attack the metabolism of birds in two ways: directly, through the nervous system, and indirectly, by interfering with the production of certain vital enzymes and hormones. The result is either death or severe calcium deficiency. The older birds having this disability are subject to nervous disorders, and may abandon or smash their eggs. (The calcium-deficient shells of the eggs are extremely thin and breakable.) If the eggs hatch, the young may die as a result of the calcium deficiency they have inherited from their parents.

Studies of the British peregrine and the American kestrel—including feeding experiments, egg injections, chick mortality counts, eggshell weight measurements, and enzyme induction—have contributed to this important discovery, which gives impetus to research in progress on the American peregrine (sometimes called the "duck hawk.")

Bird of Prey
High in a dead tree, a migrant peregrine falcon scans the Florida sky to be sure that no other hawk will disturb its meal before plucking a freshly-killed bluejay.

118

The total population of peregrines on the continent has dropped since World War II, coinciding with the spread of chlorinated insecticides. Quantities of DDT and other chemicals have been found in the eggs and fatty tissues of Western birds that nest in Baja California, Mexico and Alaska. In the eastern United States there are no longer any breeding birds available for study. Those that stream across the Gulf of Mexico each fall are migrants from the Western mountains and the Northern forests and tundras.

Tagging records have shown that the Northern birds fly directly to the Gulf Coast of Florida in a day and a half to two days, then rest and feed there before continuing on to Mexico and South America. During this interval, the tired and hungry travelers cluster on the peninsulas and offshore islands; now, more than at any other time in their lives, they are vulnerable to insecticide poisoning, parasites, and trapping. With the bulk of the North American peregrine population funneled down this way, falconers converge too. Most trap only as many birds as they need for themselves, but a few are in the hawk business. (The going price for a peregrine falcon today is $250.)

The inland peregrines occupied cliffs and tree eyries from Maine to Tennessee and wintered along the Southern coast until about 1945; then, inexplicably, they began to disappear. On their nesting grounds these falcons raised their young on small birds. Because the prey species were eating insects in DDT-sprayed orchards, much of this diet was contaminated. Now that all of the eyries (except one or two in the Adirondack Mountains of New York) are deserted, we have no way of proving that food-chain poisoning was the cause of the nesting failures. Nor do we know what happened to the adults. But the tragic fact of extinction is almost accomplished.

From the north-central section of Florida to the Keys, every slight rise in the apparently level landscape brings change. Where the soil base is

A Land of Two Seasons
In winter, south Florida's pinelands (above) *lie tinder dry and the sawgrass prairies* (below) *turn golden-ripe. Then spring and summer rains flood the landscape.*

only a few inches higher and drier, wet prairies give way to saw grass and pinelands that are flooded only during the rainy season, from June to October. At high elevations are scattered islands of trees or shrubs. Known as "hammocks," (from *hamaca*, the ancient Arawak Indian word, meaning jungle), these islands have many kinds of hardwoods, the variety depending on the location and the rainfall. Evergreen live oak, bay and myrtle flourish as far south as Miami on the Atlantic Coast, but westward toward Cape Sable the hammocks contain tropical mahogany and gumbo limbo, bare of leaves in the dry season and often growing in company with cactus.

Hammocks on the upper Gulf shelter the sabal, or cabbage palm, a tree that provides salads for people and for black bears too. Among the tropical tree islands of the south there is a scattering of tall royal palms. At first glance, one might suppose that these trees had "escaped" from the planted avenues of West Indian royal palms in Miami, but they are actually native to the area. Only a few have survived the severe hurricanes of the past five years.

The almost vertical, clifflike sides of hammocks in the Florida Everglades are sculptured by fires that race across the prairies until stopped by the permanent moats of water that usually surround them. Another kind of hammock, the cypress strand, develops in depressions that are seasonally under water. From a bird's eye view, high above the Everglades, hammocks of all sizes and shapes spatter the wide "river of grass" like ink spots. The landscape resembles a tropical savanna, and there have indeed been tropical wet and dry seasons here for thousands of years. Radiocarbon dating of peat deposits in the muddy shell marl shows evidence of fires in the "glades" for some 5,000 years. The prehistoric fires were started not only by lightning but also by Indians, who probably came into southern Florida soon after the sea retreated. Often the ancient sea basin of Lake Okeechobee, barely six feet deep, overflowed. During the heavy rains of late summer, its waters would become a torrent, sweeping a hundred miles to the salt marshes of Florida Bay and the Gulf. Then, during the winter dry season, the river would slow, the sloughs would dry, and the grass would burn.

As the climate dried and the water table fell, over the centuries, the hammocks burned too. Frank E. Egler, a close student of grassland ecology, has suggested that Indian burning may have helped to create the pattern of grass and trees that

Out of the Pouch

For more than three months, the young of the opossum—the only marsupial in North America—are nurtured in their mother's pouch. Then they progress to a place on her back, holding on like the litter shown here being transported across a Florida hammock.

exists in the Everglades today. More careless with fire than the white men who succeeded them, the Indians probably set the prairies ablaze very frequently. A fire set at the beginning of the dry season, when the grass was dry enough to burn but was still standing in a few inches of water, skimmed lightly over the glades without damaging underground roots. It stopped at the dense hammocks, some of which were under water, their foliage turgid and fire-resistant. At the end of the dry season, when the peat and trees were brittle from the drought, there was no saw-grass debris on the surface of the marshes to feed a hot, destructive fire.

By contrast, the modern glades, after years of drainage and fire protection, became a tinder box. Tons of flammable material piled up, and deer deserted such palmetto-choked pinelands as the strip from Long Pine Key to Madeira Hammock in the Everglades National Park. Sooner or later, lightning was bound to strike or a cigarette be tossed, igniting fires bigger than ever before. In 1950, the holocaust began: 23 fires blackened more than 121,000 acres of the park. A few years later, rangers started controlled burning of pinelands. They have not yet dared to burn grasslands, a step that would reduce the hazard of uncontrol-

lable fires that level hammocks and destroy grass roots. We of the generation which grew up with Smokey the Bear have nearly forgotten that fires have a place in the ecology of natural environments. For nearly half a century the United States Forest Service opposed the use of fire to clear woodlands of old, dry vegetation. And the National Park Service has been slow to admit that light and frequent fires may be necessary to preserve a park. Only recently has any progress been made in controlled burning.

At Tall Timbers (Tallahassee, Florida) and other research stations in the Southeast, however, interested groups and individuals have been at work on the problem for many years. It was discovered that too much dead litter obstructs light, intercepts rainfall, keeps seeds of pines and grasses from reaching the ground and germinating, and—by its weight—literally suppresses the upward growth of all herbaceous plants. Under these conditions, old leaves lose their vitality and dry up at the tips. After a fire, which makes room for new growth and recycles minerals in the ashes, plant ecologists have recorded 6 to 40 per cent more young green vegetation, high in protein and phosphorus content.

The research has pointed up similarities in

An Eagle's Eyrie
Perched at the top of its huge stick nest, a bald eagle looks out over pinelands on the Gulf Coast near Tampa, Florida. An eagle pair returns to the same eyrie year after year, repeatedly adding to the structure.

plant reactions to fire from Georgia to Southern Florida—and also some peculiarities. The slash pine of the Everglades, sometimes called the Cuban pine, is more fire-resistant than other pines and has its own associated ground cover. The pineland three-awn grass is especially adapted, like the dominant grasses of the tropical savannas, so that its seeds germinate better after being heat-scarred, and will send up green shoots from underground buds just days after a fire. The gallberry will sprout from underground root stalks, and the saw palmetto from underground stems, after fall or winter burning. Otherwise, these plants remain dormant during the dry season.

Among the pioneers in fire ecology were quail hunters, who gathered together in a cooperative association and proved, as early as 1928, that game birds could not scratch a living from the thick underbrush of unburned forests. In six states, from North Carolina to Mississippi, land was cleared periodically by controlled fire, and the quail returned and thrived. Herbert L. Stoddard, Sr., the coordinator of the experiments, has since disclosed that the Forest Service tried to suppress his findings and forced him to revise a book, *The Bobwhite Quail*, five times before the volume could be cleared for publication in 1931. He was only one of a number of farsighted botanists and ecologists who were maligned and threatened for holding stubbornly to the opinion that vegetation, grazing animals and game birds all benefit from periodic fires.

Quail and turkeys will flock to burned-over pinelands—almost before the smoke clears—to gobble up the pine seeds and insects that have been previously buried in dense grass. Stomachs and crops can be filled in minutes instead of hours. Occasionally a water moccasin or rattlesnake will be caught in the line of fire and killed. Nothing goes to waste, however. The dead snakes and any other sizable victims will be found and eaten by turkey vultures and black vultures, the somber sanitation crews circling overhead.

According to rough estimates from aerial photographs, about five to ten per cent of the 2,500,000 acres which comprise the Everglades—or just enough to provide water, food and shelter for wildlife in times of drought and fire—is covered by hammocks. Raccoons and opossums raise their young in the hollow trees of the hardwood hammocks, and ospreys and bald eagles nest in the tall crowns of the cypress. Neither water shortages nor hurricanes in the Everglades seem to discourage these birds, whose colonies have dwindled over much of the rest of the continent. They take their food from the Gulf of Mexico, even when fish are plentiful in the ponds and streams right below their nests. If lightning or wind destroys the nesting trees, they will raise more young another year—perhaps three nestlings instead of only one or two. Recently, 55 pairs of bald eagles were counted in the national park, and nesting success averaged 50 per cent during a five-year period (1960–1965) marked by Hurricane Donna and rainfall fifteen inches below normal.

The colonies of these fish-eating birds of prey cannot really be considered "social" groupings, but the birds here are more tolerant of other members of their own species than are most hawks and eagles. Along the Front Range near Colorado Springs, for example, golden eagles nest about five miles apart, but the records of the Park Service around Florida Bay and our own observations on the Gulf Coast north of Tampa indicate that bald eagles in Florida maintain a distance of only about a mile or a mile and a half from each other. We found nine active eagle nests in the Brooksville area by first plotting sighted birds on a coastal navigation map with a compass and then searching the interwoven woods and marshes of the peninsulas by boat and on foot. Each eyrie was within the expected distance.

For an eagle, as for a tropical fish or a garden redstart, maintaining an inviolate "headquarters," or home territory, during nesting season is absolutely necessary. This results in a fairly predictable and even distribution of eagle families over their adopted ranges. As Konrad Lorenz points out, the territory is maintained by a readiness to fight, which is greatest near the nest in the center and least at the perimeter. Aggression is affected by such factors as the stage of the breeding cycle and the time of day. Among eagles, aggression mostly takes the form of flying maneuvers and loud calling—effective bluffs that usually obviate actual fights. If a conflict develops despite all, you can predict, from the place of encounter, which bird will win: the one that is closer to its nest. When the intruder flees, its pursuer follows only a short distance into strange territory. The reason for this is that the other bird, now closer to its nest, suddenly regains the courage to launch a vigorous defensive attack. The whole performance is repeated several times until a balance of power is established and the birds threaten each other only periodically in the neutral zone between territories.

Because eagles live longer than smaller birds and occupy the same eyries for many years, the spatial relationships between them may be well established for a quarter of a century or more; they rarely, if ever, engage in the kind of squabbling among themselves that goes on between red-

starts, or even sparrow hawks. Eagles do, however, direct their aggression toward other birds. They will attack a turkey vulture that flies too low over the eyrie, and they are themselves attacked by red-shouldered hawks nesting in the same woods whenever they happen to trespass on the hawks' territory.

Presumably the hunting ranges of birds of prey out on the Gulf are free zones where anyone can fish, although the eagle has a reputation for piracy of the osprey's catch. The osprey dives deep and works hard for its living, while the eagle prefers to fish shallower water and scavenge whenever possible. The Florida eagles nest in November, and usually feed their young on Northern coots, which winter in Florida in great numbers. In some nests, however, the adults bring in nothing but big fish from the Gulf of Mexico.

Young and adult bald eagles of the Florida area may stray as far north as Canada in the summer, according to banding records, but an inland roost regularly occupied by as many as fifty juveniles has recently been discovered in the Everglades National Park, indicating that some of the birds do not leave Florida.

Nearly half of the Everglades National Park lies in tidal and subtidal flats, which are covered with water for part or all of the day. Here the tangled prop roots of the red mangrove, a freshwater plant that has gone to sea, are thickly overgrown with frilly coon oysters. Sharply spiralled tree snails crawl on the branches of the mangrove, and the unbroken canopies of dark green foliage arching overhead often bear colonies of water birds. Tides only two or three feet deep sweep the larvae of creatures from the open seas over the outermost roots; when the tide retreats, the soggy marl teems with fiddler crabs. Closer to shore, in the high tide zone, shrubby white mangrove and tall black mangrove, with pencil-thin knees similar to those of the cypress, grow on peat beds that have built up in the salt marshes. Still farther inland, buttonwood becomes estab-

A Thicket of Mangrove
In the Shark River estuary, mangroves spread over tidal mud flats in an impenetrable tangle. Aerial roots spring from each trunk just above the high tide line.

lished on the marl and other soils of the fresh-water swamp, thriving between the salt-resistant mangrove forests and the hardwood hammocks.

Red mangrove trees in abundance line the long tidal "rivers" that serrate the western rim of the glades, and their dense foliage completely over-hangs innumerable tidal creeks. The roots hold firm against the invasion of storm tides, catching all manner of debris to build up levees of marl and peat that may enclose pools of fresh rain water. The pendant green fruits of the mangrove, in which the seed germinates, drop from the trees to the forest floor, where many will become rooted. But others are swept seaward on a flood tide. After drifting with the currents for weeks or perhaps months, some of these embryonic trees may be-come stranded out in the bay, on a mud flat that is left exposed for several days by exceptionally low tides. In that brief period they will take root. If not destroyed by some later storm, the mangrove will create an island from the decaying vegeta-tion, driftwood, shells, coral fragments, uprooted sponges, and other sea growths that come to rest within their prop roots.

How much land the mangroves actually build, and where, is currently in doubt. In recent times, many mangrove keys have formed in the coastal bays of southern Florida. But the formation of such islands does not prove that the mangroves are marching into the Gulf. About 4,000 years ago the shoreline extended much farther seaward than today; according to peat samplings, man-groves grew a mile off present shores. At that time, fresh-water prairies covered areas that are now mangrove bays. But then a fairly rapid rise of three or four feet in sea level apparently turned the mangrove migration inland.

Among mangrove-destroying forces currently at work are frost, fires, and particularly hurricanes. Though the forests arch luxuriantly over islands in Whitewater Bay and the creeks of the Shark River, much of the area around Flamingo, at the eastern end of Cape Sable, has been reduced to bare marl and dead snags. The Labor Day hurri-cane of 1935 caused severe damage to these forests. In some places salt marshes replaced the shattered and uprooted mangroves; and grassy areas that were buried at that time under storm-deposited marl are still too saline for any growth. Hurricane Donna of 1960 and Hurricane Betsy of 1965 delayed by many years the possibility of recovery.

In the late summer and fall, these ghost forests and marshes seem deserted. The high ocean tides of the autumnal equinox and the rains at the peak of the season combine to inundate the coast. Then the showy water birds scatter over the southern Everglades. In winter, after the tides and the rain subside and the big inland sloughs dry up, the birds flock to the creeks and pools at Flamingo, now shallow and brimming with killifish. Sud-denly the landscape whirrs with wings as squad-rons of birds arrive and depart. So many of the herons are white or pale blue-gray, and their young look so much alike, that the easiest way to sort them out is to observe their physical adapta-tions for hunting and their special hunting tech-niques. Each occupies a special niche in a pool.

The Fishing Ponds

During January, fresh-water ponds in Everglades National Park dry up, and brackish ponds along the coast blossom with white flocks of American egrets and other water birds (left) that feed on vast schools of minnows. At this time the birds begin their spring courtships—none more spectacular than the male snowy egret's display of plumes (below) as he woos a mate.

Standing tallest, in the deeper spots, is the great blue heron. It scarcely moves until a fish swims within reach, and then it strikes with a swift and unerring bill. If the prize is small, the heron tosses and swallows it headfirst, so that the sharp fins are flattened against the body and the fish slides smoothly down the bird's long, narrow neck. If the fish is a large one, however, the great blue may walk ashore with it and beat it into the proper consistency for swallowing. The American egret, which the plume hunters called the "long white," is also a deep-wader and stalks its prey. An entire pond may be filled with these herons, mixed with lesser numbers of the snowy egret—the "short white"—whose delicate, filmy plumes blow in the wind as it scouts along the rim of the mangroves.

As mating season nears, the male snowies become quite quarrelsome, and many threat displays involving two or three birds develop. Wings spread and crests raised, the birds spar with half-open beaks until one succeeds in driving the other away. Then the fighting starts all over again, with new participants, in a mangrove alcove perhaps only a few feet away from the spot where the winner of the first "fight" is now peaceably feeding. Toward the female of his choice the snowy bows with all the grace of a Renaissance courtier, and the full courtship display is used as a greeting to his mate and young all during nesting.

There are two "night herons" in the ponds. The slender, yellow-crowned species is more likely to be seen poking about among the thickets during the daytime than the stout, black-crowned night heron. Among smaller wading birds, the Louisiana heron stays at the water's edge. The green heron, scarcely the size of a crow, often perches on a mangrove root; with a neck more than half its total length of two feet, the little fisherman reaches down into the water to catch passing minnows.

Unlike the wily herons, whose tactics are watching and waiting, the wood ibis (our only true American stork) moves along at a steady pace in the shallows, kicking up the muddy bottom with its bright pink feet and grabbing anything that comes to the surface in its scythe-shaped bill. Thousands of wood ibises nest in the maze of mangrove islands in the East River from December to June; and if there are several years of good rainfall, as many as 3,000 stork babies will hatch. But in times of drought, storms, or cold, most of the eggs and young may perish. (Closer to Flamingo, in the huge Cuthbert Lake heronry, are white ibises. Though unrelated, these smaller birds have long curved bills and habits similar to storks.) The nests of ibises and herons usually occupy the center of mangrove islands; cormorants raise their young in separate colonies on the outer fringes.

Occasionally a small group of roseate spoonbills appears in the sky, flashing scarlet wings. These big birds land in the water with a loud *whoosh*. They never fail to attract the attention of most of the bird-watchers around the ponds as they swing their strange bills from side to side, sifting rather than probing the bottom for fish, frogs and crabs.

"Is that a flamingo?" someone is bound to ask. It is a natural question, since the scarlet flamingo practically symbolizes Florida. Once the West Indian flamingo population wintered here in Florida Bay and as far north as St. John's River and Tampa Bay. But feather hunters frightened them away; by 1885, the flamingos had abandoned their northern wintering grounds. None have been observed around Cape Sable since the spring of 1902, when a flock of 500 to 1,000 left the flats, never to return. The flock at Hialeah Race Track is descended from a handful of birds imported from Cuba in the 1930's. Although nearly seven years passed before the first flamingo was born in captivity, more than 1,500 have since been hatched—and among these are the spectacular birds seen at the racing meets.

The flamingo raisers soon learned that unless the birds were fed a suitable diet of shrimps and mollusks they would lose their attractive coloring. Perhaps their selective feeding habits—as well as hunters—caused the disappearance of these birds from the wild in North America. A recent sampling of marine organisms in the pools at Flamingo turned up relatively few of the horn shells that are an important flamingo food, both here

The Spoonbill
Fishing among mangroves, a roseate spoonbill allows water to drain from its broad mandible. The bill is used to probe the muddy shallows for small marine animals.

Smallest species, the green heron, crouches in the mangrove roots and waits for a fish to approach within reach.

The Heron Family

As many as nine species of herons fish the same ponds in southern Florida, sharing a common food supply. They can live side by side almost without conflict because each occupies a separate niche defined by its physical adaptations and habits. Thus, the "little green" heron (above), too short-legged to do much wading, fishes instead from shore. The Louisiana heron (left) wades out a little way in the daytime; the yellow-crowned night heron (below) stalks the same shallows after dark. Diet varies according to size; the great blue heron (opposite page) can swallow fish weighing a pound or more and water snakes of any kind.

The yellow-crowned night heron often hunts after dark.

The Louisiana heron wades out in the water to fish.

132

Tallest of the family, the great blue heron walks knee-deep in the water and does its fishing both by day and by night.

and in the West Indies.

Flocks of young roseate spoonbills migrate seasonally from tropical America to North America. This continent always had its own spoonbill population until the era of the plume and feather trade. Then spoonbills virtually disappeared from the Texas and Louisiana coast. Most of the Florida nesting grounds, from Tampa Bay to Corkscrew Swamp and Cuthbert Lake, became deserted. In the 1920's, however, a surprising reversal occurred. Migrants from tropical Mexico reappeared in Texas and Louisiana, and many of them stayed to form new colonies in ever-growing numbers. Since the Everglades National Park was established in 1947, there has been more than a tenfold increase in the breeding population of roseate spoonbills on the keys of Florida Bay, where an estimated 350 to 500 pairs now nest. The adults probably migrate down to Cuba every year, and their young return north to the nesting area before dispersing to unknown places. An equal number of tropical spoonbills regularly appear in the Cape Sable area each winter. The relationship, if any, between the nesting birds and the migrants remains a mystery. It *is* known, largely through an ecological study by the late Robert Porter Allen, that the spoonbill here—and throughout the world —must have shallow bays, pools, and an abundance of small fishes if it is to extend its range into the temperate zone. If drainage removes its food supply or shooting threatens its survival, the species withdraws to the tropics.

Skimmers
*Black skimmers rise above Florida Bay. Though related
to the gulls and terns, skimmers have a unique,
elongated lower bill to scoop up fish from the water.*

green foliage. Though you can spot them miles away, they are practically unapproachable. Probably the bird's shyness—and a good appetite for almost any kind of food—saved the great white heron on its limited range.

"So wary are they," wrote John James Audubon in 1840, "that although they may return to the roost on the same keys, they rarely alight on the trees to which they resorted before, and if repeatedly disturbed they do not return, for many weeks at least." Audubon once raised two young great white herons himself, and was amazed that the grown birds "swallowed a bucketful of mullets in a few minutes, each devouring a gallon of these fishes." The birds also killed some young reddish egrets and Louisiana herons and swallowed them whole, "although they were abundantly fed on the flesh of green turtles."

Florida Bay is a world apart from the bird ponds at Flamingo. The pulsing tide of a few inches floods the shoals almost daily. Then black skimmers rise from shore roosts in unbelievable numbers to forage for fish and shrimp beneath the calm surface. Their half-open bills cut the water like scissors. When the long lower mandible strikes a fish, the upper bill clamps down tightly, and the bird swallows its prey without missing a wingbeat. With the waning tide, many square miles of the muddy bottom are exposed, leaving an intricate maze of winding channels, and these now become the hunting grounds of herons, gulls and vultures.

The sea-grass beds shelter sea turtles and the rare manatees, or sea cows. Southern Florida is probably the last refuge of this marine mammal, which is hunted in tropical Australia, Africa and South America for its meat, oil and commercially valuable hide. Scientists at Florida Atlantic University believe they have found a way to use the sea cow and at the same time save it from extinction. One adult manatee may eat a ton of aquatic vegetation a day. Since many estuarine rivers and canals in the warm temperate zones of the world are now choked with water hyacinths, intro-

Another victim of the plume hunters, the reddish egret, has survived in the United States only in southern Florida. Fewer than 300 of these birds which nest in Florida Bay represent practically all that is left of their once sizable native population. The egrets, varying in color from white to rusty, cluster in the shallow waters around the keys or far out on the sand shoals and mud banks of lagoons. There they stand motionless, or walk about slowly in search of prey until the rising tide forces them to take flight.

The great white herons have never been known to live anywhere except on the keys of Florida Bay and on the outer coast. When not fishing, they perch in small groups on the red mangroves, their pure white plumage contrasting with the dark

duced from the tropics, manatees could be a boon in clearing channels for navigation. Chemical and mechanical controls eliminate the hyacinths for only one to three months, but a hungry sea cow uproots entire plants and prevents new growth for six to eight months. The only problem in Florida is the vulnerability of this tropical animal to cold. If air and water temperatures drop, many individuals die of respiratory diseases and bronchial pneumonia.

As the concern over disappearing species grows, we are beginning to realize that the reasons for the disappearances are complex. Some species, like the great white heron, are more adaptable than others. If we want to ensure the survival of the less adaptable ones, we may have to raise them ourselves. This is exactly what the Bureau of Sport Fisheries and Wildlife has in mind. At Laurel, Maryland, this federal agency has a "nature bank" of rare birds and animals that may eventually provide new stock—and new solutions for problems of survival in the wild. One without the other is useless. For example, the Everglade kite long ago disappeared from the Everglades, and now it breeds only around Lake Okeechobee and the Loxahatchee Wildlife Refuge southeast of the lake. The Bureau is studying an Argentine race of kites in an attempt to discover what is happening to the Florida kite population, now down to seven or eight pairs. It appears that drainage has either destroyed the snails on which the kite feeds or has concentrated them in ditches, thus increasing the incidence of a lung fluke carried by the snails. It may be that this fluke has decimated the birds. Six years ago, we conducted postmortems on several dead kites and found them to be heavily infested with the parasites.

Today the Everglades is no longer a natural river. Much of the land above the park boundary has been altered by water-management programs. Some 44 per cent—1,208 square miles—has been drained for farms or set aside to be developed as farms and residential areas. Another 49 per cent —1,345 square miles—consists of three conservation areas under the joint control of the Central and Southern Florida Flood Control District and the Army Corps of Engineers. Of the entire "river of grass," only seven per cent falls within the Everglades National Park that has been estab-

lished for the preservation of the natural habitat.

In 1906 the first levees were erected to hold back seasonal flooding from Lake Okeechobee. Despite these "flood protection" works, hurricanes in 1926 and 1928 scooped the water out of the shallow basin, as though from a saucer, and several thousand people were drowned. By 1936, sixteen million dollars had been spent on gates and levees. Before the hurricane season every year billions of gallons of water were released to the Gulf and to the Atlantic through the Caloosahatchee River and the St. Lucie Canal. During the dry season, however, the uncoordinated canal system could not keep farmlands from drying out, and the surface of the soil actually sank as much as eight feet in some places. In the rainy season, flooding continued. Meanwhile the growing resorts of the east coast were pumping increasing amounts of sweet water from wells. The water table dropped, and the sea invaded the porous limestone aquifer around Miami. Brackish water flowed from city taps.

To raise the water level and keep it stable, the Corps of Engineers and the Flood Control District collaborated in the construction of a vast system of shallow, grassy reservoirs. Open to the public, the reservoir areas provided fishing, camping, boating and other recreational facilities, but their maintenance was to have a drastic effect on the new national park to the south. After the engineers completed Area 3, just north of the Tamiami Trail in 1962, the levee and the new road that had been constructed effectively sealed off the lower glades, except for six sets of gates—four to release water into the park, and two used as emergency outlets from the conservation area. During the next four years, a period of very low rainfall, little or no water was released through the gates. Because there had been no provision in the plans to carry the overflow from Okeechobee through the drained farmlands, the park nearly died of thirst. The imperceptibly moving river, which in the past had been at least three feet deep at the beginning of the winter dry season, dropped to a few inches around the grass roots, and sloughs that had never gone dry were parched by April or May.

In May of 1962, the only movement of water through the park was seepage under the levees. Both the quantity and quality of the water

Before and After

On Florida's Gulf Coast, palm and wiregrass marshes (above) are converted by real estate speculators into deserts (below) filled with crushed limestone dredged from beneath the tidelands. Hundreds of square miles of this shore lie barren and uninhabitable for wildlife—and for people, too, until lots can be developed.

changed. In an area of 75 square miles, the water level fell 1.5 feet below mean sea level. As the normal flow of fresh water to the coast stopped, the salinity of shallow bays increased, endangering the nursery grounds of many marine fishes and crustaceans, notably the pink shrimp. In April of 1965 the crisis reached a peak. By now, even the crayfish burrows and 'gator holes in the Everglades were dry. One of the park naturalists pumped water from a 30-foot well in an attempt to save the fish. Rangers roped alligators and transported them by truck to distant water holes. The alligator-roping received much publicity, but the threat to city drinking water and to sport and commercial fishing was really of greater concern. For the first time, everyone—naturalists, engineers and fishermen included—realized that more than a lovely park was at stake. They began to *want* to understand the complex hydrology that links the glades to the sea coast in an interdependent relationship.

During the spring of 1966, the Flood Control District released 130 million gallons of water a day, enough to keep some water in all the main sloughs. The following year, canals connecting Lake Okeechobee with the conservation areas were enlarged to provide a greater and more rapid flow to the south, aided by pumping. Meanwhile, the weather changed; heavy rains came for two consecutive years, filling southern Florida with water again. The next spring, a sudden release of water by district authorities caused the accidental drowning of many deer. Water in southern Florida continues to represent a problem of too much or too little.

Recently the National Park Service has come under a good deal of criticism for dredging a new, 120-mile-long canal from Flamingo into the mangrove swamps of the Gulf Coast, for the recreational use of small-boat owners. Fresh water flows out too fast and salt water penetrates too far inshore, changing the ecological balance of brackish lagoons. The natural consequences of such dredging may be far worse than the possible loss of a few tourists. Here, recreational development conflicts with the preservation of one of our last truly wild coastlines.

All along the Gulf Coast, massive dredging and filling for housing developments has destroyed nursery areas vital to the production of fish and shellfish. Instead of marshes, there are vast wastelands of fill, waiting for houses and landscaped lawns. Florida's Bulkhead Act, passed in 1957, has set some limits to coastal development, requiring that sea walls run no farther into the Gulf than mean low tide, and that they follow the natural curves of the shoreline. The Act specifies that offshore bars, grasses and mangrove islands shall be left intact wherever possible.

In many ways a unique example of cooperation between developers and wildlife agencies is Marco Island, the "gateway to the 10,000 Islands, to the Everglades National Park, and to Shark River." The community has the familiar Florida look—a canal at every doorway—but the builders here devised a method of "digging in the dry"; that is, they diked large areas first and lowered the water table by pumping. Only after the fill was completed and the sea walls were installed did they open the canals. Every bald eagle nest has been allotted a two-acre no-trespassing zone, and the entire island is a Florida Audubon Society Sanctuary for the national bird. In addition, the Society owns and administers a 1,600-acre refuge in the mangrove-studded bay, the home of thousands of roseate spoonbills and other wading birds and divers.

Another encouraging precedent in the line of wildlife preservation was the establishment in 1966 of a 10,000-acre preserve in Estero Bay, near Fort Meyers—the first of six planned by the Florida Outdoor Recreational Development Council. With only twenty per cent of the State's 8,000-mile coastline developed, there are still chances to plan a diversified landscape in which roseate spoonbills, eagles and people can live together, sharing the same rich, estuarine fishing grounds.

4
Upland and Northern Forests

Northern Coniferous Forest

Rain Forest

Redwoods

Mixed Coniferous
and Deciduous Forest

Many centuries ago, the Anglo-Saxons called the forests that covered England the *wilddeoren,* or "place where the wild beasts live." Until the Middle Ages, the greater part of western and central Europe was one immense forest, broken only by heaths and swamps. From 1050 A.D. on, however, the Old World was intensively cut over, cultivated and grazed by domestic animals; and the wilderness almost disappeared. Coming from highly developed lands, the first European explorers in North America were amazed and often overwhelmed at the sight of magnificent "virgin" forests filled with creatures that had never known the yoke of man. Verrazano, Cartier and Champlain among others, thought that they had found a wilderness untouched since the creation.

The truth was far different. A relatively small population of Indians had been chopping and burning the American forests for thousands of years. Though limited in technology, they were capable of causing great changes in the landscape. Some ecologists believe that fires set by nomadic hunters of bison in our Middle West helped to maintain prairies in regions that are forested today. Only after Iowa, Illinois, Indiana and Ohio were settled, and Indian burning stopped, did the trees return. Two hundred years from now, the silver maples that took root along Midwestern streams after 1830 may be considered "virgin."

Because of the size of the United States, and its diversity of habitats, about four times as many species of trees grow here as in all of Europe. Covering the 500-million-year-old mountains and plateaus of the American East and the relatively newer mountain areas of the West are many different forest communities.

The growth of forests is limited by temperature and rainfall; the variety of trees is determined by local variations in the composition and microclimate of the soil. Largest of the tree communities on the North American continent is the Eastern temperate forest, a mixture of deciduous and coniferous trees growing in an area which enjoys 28 to 60 inches of precipitation a year. Oaks, maples and other trees that drop their leaves in winter predominate throughout the Middle Atlantic states and New England. As the climate cools to the north, conifers with flattened, needle-shaped leaves, resistant to cold and drought, outnumber the deciduous trees—as they do on the higher slopes of the Blue Ridge and Great Smoky Mountains to the southward.

In the fog and drenching rains of the Smokies, which may receive up to 100 inches of rainfall a year, the individual trees are larger and the forest denser than elsewhere in the East. As many as 20 different kinds of trees reach record proportions in cove forests here that are rivaled only by the Pacific rain forests, which may have 130 inches of rainfall a year. Neither of these regions was disturbed by the continental glaciers of the Ice Ages. Throughout their long development, abundant flowing water has dissolved minerals from decaying forest litter and promoted the growth of many-layered plant communities. No stone is left uncovered by moss. No fallen log is without a companion group of young trees, trying to succeed under the dark canopy of foliage overhead.

The character of the Eastern forest has been changed more than any other since the early days of this country. By 1830, from 80 to 90 per cent of New England had been converted to farms. But the crop yields were low and the work was hard. When the prairie region to the west opened up, offering deeper and richer soil, the New England hills were largely abandoned. Today, 60 per cent of the New England forest has grown back. However, the second growth forest is not the same as that of, say, 1500.

The classical theory of plant succession indicates that where deciduous trees have grown, they will one day return, but the United States has not been settled long enough to provide proof of the theory. Instead, we observe patterns of regrowth that vary from place to place. In Salisbury, Con-

necticut, for example, a deserted field is first filled with goldenrod and hard-hack. Then, pushing up through the weeds, come the scrubbier trees—gray birch and white pine, which foresters call "volunteers" because they seed profusely and grow in the open. Young pines, however, cannot survive in the shade of old pines. The red maple does thrive in shade, and this tree takes over the growing space—later to be joined or supplanted by the sugar maple, hemlock, beech, yellow birch and various oaks and ashes.

That is the situation in Salisbury. But elsewhere in Connecticut, dissimilar patterns have persisted for many years. There are areas in which oak forests have sprung directly from old stumps, giving pines and maples little chance to grow up. There are logged and burned-over pinelands, on which only the most fire-resistant trees, such as aspen and birch, are left. And there are shrubby patches with no trees at all.

Conditions in the various sections of regrowth have benefited opossums, raccoons, deer, song birds and other wildlife of the forest edges. In the old forest, deer were less plentiful because they had less new growth to browse on. Now, for several decades past they have bred fast, and their numbers have shot up—with lasting effects on the second growth forest. By browsing down oak and maple, for example, the deer of Salisbury

Township have given beech seedlings a competitive advantage over the other deciduous trees. (An overpopulation of deer in the Northeast results also from the extermination of the cougar and the wolf in the region, leaving no natural predators to check the increase of the herds.)

When the original American forest was cut down, for lumber and to create farms, the wildlife that had depended on the old forest for shelter and food either declined or vanished. By 1800, most of the Eastern woodlands had been cut over. Carolina parakeets and passenger pigeons became concentrated in a few enormous nesting colonies, where they could be easily trapped or gunned down. Soon the spectacular parakeets were gone. The passenger pigeons survived only in the beech forests of the Midwest. In 1870, one of their final nesting places was reported to cover hundreds of square miles of forest in Wisconsin, each tree having dozens of nests. As the beech trees fell before the lumberman's axe, the pigeons became fewer and fewer. Only thirty years later, these birds that once numbered in the millions had disappeared entirely. By a strange coincidence, the last captive passenger pigeon died in a zoo in 1914, the same year that the last captive parakeet died. For both species, loss of habitat had led directly to extinction.

The "conquest" of the old wilderness undoubtedly bettered the material life of the pioneers. But some of their descendants are now wondering whether the rewards received compensate for the loss of phalanxes of native pigeons in the spring, or such delights as the "grassy glades" and "very good woodland . . . with fresh water running through . . . pleasant and proper for man and beast to drink" that was Manhattan as Henry Hudson described it in 1609. Today suburbanites can enjoy the scrub forests on old farmsites surrounding the city, but the waters are for the most part impure, and fish suffocate in the wastes that the rivers are unable to carry to the sea.

Westerners, who live in a part of the country that was settled later and is still relatively unpopulated, have less reason to mourn the loss of "virgin" woodland. The Northwestern coastal forests, growing in a region of high rainfall and fog, without severe winter cold, produce more timber per acre than almost any other forest area on earth. Land lumbered only a hundred years ago is completely covered now with young, vigorous trees. The California redwoods, in particular, show remarkable resilience; for at least 10,000 years they have resisted flood and fire. And today the redwood forests are in no danger of becoming extinct. Fortunately for all Americans, thousands of square miles of Western timber have been set aside in state and federal preserves. East of the Mississippi, no living man will ever see the "forest primeval," except in a few plots of a thousand acres or less.

Beyond the point where the granitic rocks of the Appalachians drop off, to be replaced by the sedimentary rocks of the coastal plain—a line marked by waterfalls (and major cities) from the Great Falls of the Potomac above Washington to the falls of the Chattahoochee River at Columbus, Georgia—the Eastern deciduous forest reaches in long fingers to the sea along river courses, through surrounding pine barrens and salt marshes. On an island in the Patuxent River near Laurel, Maryland, there is a seven-acre stand of beech trees which escaped colonial logging and is now part of a 2,670-acre preserve. Like other remnants of "virgin" woods in the East, the tract is too small to support all of the life that was there in 1500. The tracks on the bank of the river are left only by deer and raccoon; there is no trace of a wolf or cougar. In the spring, the big trees tower over a ground cover of rare dwarf ginseng, the delicate white flower whose relatives in the Orient are believed to hold the secret of eternal youth.

The oak, beech and maple trees of the Patuxent forest have a life expectancy of at least 300 years. During this period the trees will maintain the kind of soil necessary for their growth—as well as a microclimate under the canopy that is darker, calmer, and less prone to sudden temperature changes than is to be found in open country. With the advent of winter, their period of rest, the trees shed their dead leaves. The leaves decay,

The Forest Canopy
Crowns of old beech trees at the Patuxent Wildlife Research Center in Laurel, Maryland, interlock in a lacy pattern of sunlight and new spring leaves.

Running Deer

Twigs snap beneath its hooves as a white-tailed buck bounds over the brushy floor of a pine-oak woodland. When alarmed, the deer "flags" its companions by raising the white underside of its tail, and dashes for safety deep in a thicket.

Climbing Coon

After a swim, a wet raccoon climbs to its observation post overlooking a forest stream. During the spring, the coon may be found here early in the day and again in midafternoon, patiently watching for a fish to appear in the clear water below.

and return a rich, carbonaceous layer of fertility to the soil. When their arboreal plumbing systems are turned on in the spring, the trees leaf again, to become the most efficient of humidifiers. Water piped up from deep tap roots is transpired, or breathed out, through the pores of the leaves in quantities of millions of gallons every day.

Forest herbs vary from a few inches to several feet in height. Shrubs up to fifteen feet high shade the herbs, and trees that may reach to 35 feet or more overtop all other woody vegetation. When tall oaks predominate in a forest, they cut out 90 per cent of the sun's light, and only shade-loving plants become established in the understory. Among these are red (or swamp) maples, hickories, spicebush, nettles, bloodroot and ginger. Below ground, the roots of plants compete at different levels for water and nutrients.

In this crowded world, the animal life sorts itself out into various strata. The soil basement is tunneled by worms, shrews and moles, and within leaf mold the eggs and larvae of insects are incubated. Many of these insects are only transients on the forest floor; for example, certain leaf-eating beetles and "ladybirds" that prey on aphids mi-

grate seasonally to higher niches. In general, the parasites and predators are more mobile than the great mass of herbivorous insects, which avoid competition for food by staying put. There are aphids, scale insects, and borers that continue to lay their eggs, generation after generation, on certain plants—or even parts of the same plant— where the parent insects fed during the larval stages of life. This kind of genetic programming tends to keep them at a particular level of the forest. From subterranean roots to tree tops, each different level has a wide assortment of specialists, none conflicting with the others.

A leaf may be attacked by insects in many ways. There are caterpillars that scrape the surface, or tunnel between the leaf's epidermal layers, or roll the whole thing up and eat it. Other insects suck the leaf sap; still others cut out portions of the leaf with specially adapted mouth parts.

Among the mammals that live continuously on the forest floor are deer, which can eat only the herbs and shrubs within reach. Chipmunks and deer mice den at the bases of trees, but climb and forage in the understory. Raccoons use fallen logs as roadways, and raise their young in the hollow

147

stubs of dead trunks. Being omnivorous, they may harvest the nuts and fruits of the forest for a meal or dig up a nest of mice, depending on the season.

Though birds have the freedom of flight, they are often limited to narrow niches in the forest habitat. Food preferences are important, but not always the main reason why one species of bird spends more time on a given level of the forest than another. The flycatcher, to be sure, stays in the treetops because of its hunting habits. But for the woodpecker, a nest hole high in the middle to upper levels of the tree trunks is what counts the most. Vireos and chickadees have favorite song perches, from which they do most of their singing.

Having a niche of one's own reduces competition among close relatives with similar feeding habits. As many as eleven species of warblers may occupy different strata in a mixed coniferous and deciduous woods. The Blackburnian warbler is found at the top levels of spruce, fir or hemlock, and the black-throated green warbler just below.

The magnolia warbler occupies a lower level among the evergreens. The redstart flashes a brilliant tail (the badge of the male) as it dashes after flying insects in the understory of maple, ash or other deciduous trees. The black-and-white warbler canvasses the tree trunks for crawling insects. four other species do their feeding in sunlit or shaded shrubs, sometimes descending to the forest floor. At the very bottom, the ovenbird and the Nashville warbler are almost completely terrestrial; they nest and scratch their living out of the forest carpet of old leaves. In a deep leaf bed the ovenbird builds an elaborate home, roofed over with a variety of finely woven materials. The resemblance of this nest to an outdoor oven gave the ovenbird its name. Fallen branches and leaves ordinarily conceal the structure so effectively that it is almost impossible to find except by accidentally flushing the bird.

The niche of a large predatory bird, such as a hawk or an owl, has wider dimensions. The red-shouldered hawk builds a nest in the tree tops,

A Leafy Bed

Camouflaged by leaves on the forest floor, an ovenbird quietly incubates her eggs. After the young hatch, the parents have been observed to bring insect food for them a hundred or more times a day.

The Highest Cradle
One hundred and ten feet above the ground, a week-old red-shouldered hawk nestles among fresh beech leaves. Its parents supply this nest lining, which is changed frequently.

and it has four or five favorite roosts, both at home and along its accustomed route to hunting areas. These roosts are at many different heights in the trees, and are used for several purposes: to watch for climbing enemies like the raccoon, to spot prey, or simply to rest and preen. The hawk feeds mostly on amphibians and reptiles in the marshes and along the banks of streams. On these hunting grounds the red-shoulder has no competitor except the barred owl, a nocturnal fisherman that hunts only after dark, when the hawk is asleep at its roost. There are differences in hunting tactics, too. Unlike the hawk, the owl actually dives into the water to capture a frog or a snake, and it consumes many fish.

When observed as a living community, the forest is built in all its complexity upon the smallest of organisms. Whatever happens at the microscopic level of bacteria, fungi, and viruses is sure to affect the large predators at the ends of food chains. The abundance of the red fox, for example, is in part determined by the prevalence of a particular insect disease. The fox eats cottontail rabbits, which are nourished by cabbage-like plants. These plants can be destroyed by the cabbage butterfly, which in turn is attacked by a parasitic wasp, which is susceptible to a disease-producing microorganism.

Looking at forest life in another way, ecologists recognize that there is a "feedback" from the ends of food chains. Thus, predation by hawks, owls and foxes initially increases the mortality rate of rabbits above the birth rate. Then the rabbit population levels off, because the surviving individuals are better fed and live longer. They are also less vulnerable to attack; a small number of animals can hide more effectively under plant cover than many. It is very difficult for the predator in nature to destroy its own food supply. Only man is capable of killing to the point of extinction. He kills for food, for sport, and for revenge—and, in doing so, he often succeeds in upsetting the delicate adjustments that have evolved over thousands of years between plant and animal, parasite and host, predator and prey.

Since man has killed off the biggest carnivores, a very important ecological relationship has disappeared from the Eastern forests. It has been estimated that the white-tailed deer population varied from about 100 to 800 for every ten square miles of woodland in colonial days. One or two wolves or cougars in the area kept the numbers of deer stable by killing about a quarter of them each year. But as the land was settled and the wilderness shrank, the predators vanished. Cougars and wolves were rapidly shot out. Bobcats, which prey on very young, sick or old deer, could not replace the larger predators in holding down the herds, nor could the omnivorous black bears, foxes, raccoons, skunks and opossums. The human hunter, for his part, was comparatively inefficient; and anyway, he wanted more deer to hunt. Game was protected with closed seasons as early as 1677 in Connecticut, and there were bounties placed on cougars and wolves as early as Plymouth Colony's founding. In consequence, the deer multiplied unchecked, feeding on the scrub of hill farms and lumbered-off forests.

By 1880 there were already too many deer on Mt. Desert Island, Maine; six years later, during winter ice storms, the deer met with wholesale starvation. The Adirondacks experienced their first deer irruption in 1896, about a decade after the wolves and cougars were gone. The pattern of periodically soaring deer populations continues into the present, with harmful effects on the forests and on the deer themselves.

In a woods heavily browsed by deer, only the less palatable plants and leaves will be left to renew themselves. Tulip trees, among other species, cannot seed successfully; the seedlings are eaten before they are three months old, and are found only where they have sprouted from stumps. Hickory nuts are particularly relished by deer (as well as bear), and in some cases the entire crop of nuts is consumed before any can germinate. A game manager can tell by the browse line on flowering dogwood, laurel and rhododendron—often clipped down to the stumps—just how serious the deer increase has become. Next to go are the lower branches of trees, four or five feet off the ground.

Now the deer must feed on less substantial browse. Although the animals may still appear well fed, they gradually become victims of malnutrition and disease, and finally starve.

By killing the predators and protecting the prey, man has started a destructive and perhaps irreversible cycle in which large herbivores can and often do destroy their food supply. Even if there were more deer hunters in the Eastern forest, a massive kill by expert marksmen would not accomplish the same result as the predator's steady, year by year culling of the old, the young, the vagrant and the ill-adapted animals from the herd. The cougar and the wolf left a nucleus of healthy, well-adapted deer to breed and produce healthy fawns. After almost a century of freedom from natural predation, Eastern white-tails are for the most part in poor condition.

In a dark, old forest, the death of a giant tree opens up the canopy, making possible new life in the sun—new pastures for rabbits and white-tailed deer. The fallen tree trunk is bored into by beetles, and also by birds looking for beetles. Under the loosened bark a whole colony of larvae and mites collect, to be followed by their predators, the carabid beetles. Then shelf fungi soften the sapwood, and their mycelia become thickly extended through the bark. Wood-eating termites and carpenter ants move in, together with at least 50 more insect species that live on the fungi. Finally the heartwood is reduced to a crumbled mass of soil by red rot fungi. Lichens, mosses and saprophytes such as the pale Indian pipe, fed through a liaison with fungus roots, are pioneers in this new soil, which eventually becomes a seedbed of organic richness for higher plants. From dry fallen leaves, earthworms manufacture castings rich in phosphorus, nitrogen and potassium.

A familiar gray fungus takes over a stump.   *Lichens display a rich variety of color.*   *Pinedrops nod their buds.*

A Log and Its Destroyers

Felled by wind, lightning, or old age, a tree will gradually crumble away
and return to the soil. Among its destroyers are plants without chlorophyll,
which cannot manufacture their own food from sunlight.
The shelf fungi (above, left) are the fruits of threadlike mycelia
that invade and break down wood. Other fungus threads surround algae
and form lichens (center). In a symbiotic arrangement, the fungus receives
moisture and food from the green plant, and the algae are given anchorage
on the decaying log by their rootlike fungus partners.
The budding pinedrops (right) is a saprophyte, unable to live
unless its roots are entwined with fungi that cause fallen plant material
to decay and thus change into usable food for the forest community.

Struck down by lightning, a big old tree lies on the floor of an Eastern forest. Soon the process of decay will begin.

Their myriad tunnels aerate the topsoil, and their plowing spreads its mineral content, extending the depth to which roots can grow.

Seventy-five years ago a German scientist looking for truffles discovered that fungus roots are important partners of deciduous trees—oaks, beeches, chestnuts, alders and hornbeams. The stubby rootlets of these trees are in each case embedded in a layer of fungus. By secretions of certain mutually beneficial chemicals (auxins), the roots and the fungus—the latter an efficient gatherer of nutriments—stimulate each other's growth.

In the forest of the Canadian North, the overall population of decay organisms and earthworms is low, because neither can thrive in the highly acid, gravelly soil of this region. But the fungus partners of the spruce and fir are specially adapted to the acidity, and these conifers absorb up to 85 per cent more nitrogen, 75 per cent more potassium and 234 per cent more phosphorus than other trees. Flowers are scarce here except for miniature groves of saprophytes and a few scattered orchids. Like the Indian pipe and its relatives, the orchid depends on fungus roots for its life. Its tiny seeds, that do not contain any stored sugar, cannot survive long and will never germinate unless they are invaded by the right kind of fungus. Consequently, the plants are sporadically distributed; in a day's search through the spring woods, you might come across only one or two lady slippers—or you might encounter a mass of them. New species of orchids are sometimes discovered and then lost again for several decades before their rediscovery.

Spruce and fir, dominant in the Canadian forests, are also much in evidence along the jutting headlands of the coast of Maine, where cold sea winds and salt spray exclude other trees. The floor of an evergreen woods is a spongy carpet of moss-covered needles. Instead of hundred-foot trees of great girth, as in old deciduous forests, there are slim trees scarcely fifty feet tall, in dense and uniform stands.

These even-aged stands are the product of severe fires that periodically sweep the landscape, destroying thousands of acres of trees. The regeneration of the forest begins with a low ground cover of bushes, usually succeeded by trees that sprout from burned-over roots and stumps—birch, pin cherry, aspen, sumac and pitch pine. The seedlings of the spruce and fir are present, but are suppressed in the shaded understory. When the short-lived "fire trees" die off, admitting sunlight to the forest floor, the conifers will quickly spring up to achieve dominance.

This is a resilient but unstable environment. Like a field of corn, a spruce stand is a monoculture. If weather conditions are favorable and the crop is good, there are great irruptions of spruce beetles and budworms. The most serious pest, the spruce budworm, irrupts every thirty-seven years, and kills vast tracts of balsam fir and red spruce. The budworm moth lays its eggs in the tops of the conifers in July. On hatching, the caterpillars spin silken cases in which they pass the winter. Then, as new needles open in the spring, they emerge and spend three to five weeks consuming the tree crowns. The trees can stand only two defoliations before they turn brown, as though scorched. Subsequently, birds flock in to eat the marauding insect hordes. All the available nesting places are filled, and each mated pair produces four to six more eggs than usual. But neither the birds nor any other predators are able to control the insect outbreak.

By cutting out the spruce in many parts of the maritime provinces of Canada, the lumberman has contributed to the insect plagues which have swept the North repeatedly in the last century, each onset said to be worse than the last. The "spruce" budworm is really an inhabitant of balsam fir stands rather than of spruce. The caterpillars develop and feed most abundantly on the needles of the fir and the pollen of its tiny staminate flowers. This fundamental information about the insect was discovered long after conversion of the forest to a community in which balsam fir predominates. And now the budworm is practically impossible to rout.

During the last decade, foresters have tried to restore the balance of the forests by cutting out balsam fir—especially older stands that have less resistance to insect attack and disease than

More Rocks than Earth
On Mount Desert Island, Maine, the shallow roots of spruce trees spread over boulders. Only trees that do not have taproots can thrive in such meager soil.

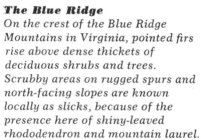

The Blue Ridge
*On the crest of the Blue Ridge
Mountains in Virginia, pointed firs
rise above dense thickets of
deciduous shrubs and trees.
Scrubby areas on rugged spurs and
north-facing slopes are known
locally as slicks, because of the
presence here of shiny-leaved
rhododendron and mountain laurel.*

younger ones. As a last resort, DDT was sprayed over many millions of acres in New Brunswick; between 1952 and 1959 a number of test areas were sprayed at least twice. After each treatment insect numbers were temporarily curbed, but unexpected damage was done to other forest life. Thus, in the treated tracts only half as many young woodcocks hatched as in the untreated ones. Runoff of the insecticide into the Miramichi River killed susceptible young salmon and the insect food of many adult salmon, causing the latter to die of starvation. Entire generations of fish were lost.

Among the uncounted dead were many small warblers that ate DDT-poisoned larvae. But the demise of these birds was hardly noticed because of the steady stream of new arrivals from other areas, drawn by the insect abundance.

The spraying operations stopped when the insect infestations subsided. But probably these irruptions would have subsided anyway, without the aid of modern technology. Ironically, the use of DDT wiped out some of the natural predators of the budworm larvae. Canadian biologists found that web-spinning spiders had been capable of destroying many budworm eggs, and these spiders suffered heavily from the insecticide. The next budworm outbreak may be more severe; if the usual predators and parasites of the insect are gone, or greatly reduced, the new generations of budworms will have a better chance of survival —until man's chemical warfare begins again.

The southernmost extent of so-called "Northern" conifers in the Eastern United States is the thin remnant of spruce and fir along the tops of the Appalachian ridges. You can still travel the whole length of this mountain range on foot by way of the Appalachian Trail. The journey will cover some 2,000 miles and—according to a hiker who recently walked from Georgia to Maine—will take four months. Automobile tourists prefer the 469-mile Blue Ridge Parkway, which follows along the southern mountain crests in linking Shenandoah National Park in northern Virginia to the Great Smoky Mountains National Park in North Carolina and Tennessee.

Below the fir, in the 1,500 to 4,000-foot zone,

A Home in a Tree
Among many familiar birds that customarily inhabit the edge area of a forest is the bluebird. Here a mother bird arrives at the rim of her nest hole in an old apple tree, bringing a spider for her nestlings, one of which can be seen begging for this choice morsel with wide yellow gape.

immense forests of American chestnuts have been wiped out by a blight introduced from the Orient, when Chinese chestnut lumber was imported around the turn of the century. The fungus disease was discovered on native chestnut trees in the New York Zoological Park in 1904, and within a few years it had spread throughout the Northeast. By 1935, the blight had swept south to Harlan, Kentucky. Soon the nut itself, which Champlain first tasted three hundred years ago in the St. Lawrence River Valley (and praised as the sweetest he had ever eaten) was practically gone from North America. Although green sprouts still grow from some of the old stumps, none of the trees ever lives more than a few years. The fungus attacks and kills them. Today, in dead chestnut groves the great trees stand stark as grave-markers, or else lie in jumbled heaps on the ground. Rain has seeped into the chinks and knotholes, and the insides of the big logs have rotted away, providing warm hollows in which animals hibernate through the winter. Among the hibernators is the broad-headed skink, a Southern amphibian that ranges as far north as Maryland. Here, during cold weather, skinks can only be found in chestnut hollows. After all the old logs have crumbled away, the animals may not be able to survive in the Middle Atlantic region.

Gradually the gaps in the mountain forests left by the departed chestnuts are being filled with other trees, and small birds and deer are returning. Along the Blue Ridge, new young forests of red oak, chestnut oak and white oak have sprung up. In the Great Smoky Mountains, where chestnuts once made up 50 to 80 per cent of the canopy, there are now dense stands of chestnut oak, red maple and sour wood.

In the Blue Ridge country of early days, even the most remote areas were reached by the mountaineer with his squirrel rifle. So many wild animals, both large and small, were killed that several species disappeared from the region forever. The Eastern elk, or wapiti, was extinct by the middle of the 19th Century. Except in scattered hemlock coves, where trees over 400 years old and as much as three feet in diameter still stand, the forests of 300 years ago were cut for wood products and

The mother bluebird brings home a moth.

A caterpillar provides a juicy tidbit.

Home Life

*The domestic life of a pair of bluebirds is
hectic in the extreme until the young are graduated
from the nest, about ten days after hatching.
The parent birds hunt the air, the foliage of
the trees, and the open fields for enough food to
satisfy their ever-hungry nestlings. While the
mother brings home insects, the male (below) may
busy himself with cleaning the nest hole of waste
matter, which the young birds deposit in fecal sacs,
the avian equivalent of disposable diaper bags.*

The male bird takes away a fecal sac.

A High Apartment

Flashing a red crown, a pileated woodpecker arrives at its doorstep, about sixty feet above the ground. This bird, as large as a crow, needs a big tree in which to chisel its nest hole. It also needs an extensive supply of bark beetles and big black carpenter ants, which are favored dishes of young pileated woodpeckers.

to clear the land. Regrowth of the oaks and other trees was prevented by fire and by grazing, With wildlife depleted and forests gone, the mountain people for the most part migrated West, or down to the cities in the valleys.

In 1925, the State of Virginia began acquiring some 3,870 private tracts which would become the Shenandoah National Park, dedicated in 1936. Spindly, second-growth stands of timber are still reaching for their full size here today. The cabins of one-time settlers in the area are relics to be viewed from Skyline Drive, and all of the formerly cultivated fields—like Big Meadow—are dotted with scrub pine, black locust, sassafras and other invaders, signaling the return of the deciduous forest.

As the landscape changes, so does the bird population. The Eastern meadowlark leaves the meadow when that place becomes weedy enough to suit the grasshopper sparrow, the Henslow's sparrow, and the indigo bunting. When growth reaches the shrub stage, the former meadow is favored by song sparrows; then the shrubbery is gradually taken over by towhees and brown thrashers. Yellow-throats nest in a moist tangle of weeds and briars. Some years hence, the pines that are now only waist high in Big Meadow will be large enough to attract redstarts, bluebirds, oven-birds and thrushes. If there is no fire or other disturbance, oak trees may grow here within the

next century, and they will probably house scarlet tanagers and pileated woodpeckers, the birds of the original stands.

The preferences of certain birds for certain habitats are very often due to long-established behavior patterns. The tanager, for example, needs a firm anchorage at the end of a tree branch for its cup-shaped nest; the woodpecker, a tree trunk large enough to carve out a nest hole; and the bluebird, a deserted hole or natural hollow. Some of the smaller birds of the fields seem to demand water nearby. Ornithologist Joseph Hickey once decided to test the rule that song sparrows and yellowthroats always nest close to a pond or stream. He drove through 5,000 miles of the Wisconsin hills before he found a song sparrow more than 50 yards or a yellowthroat more than 150 yards away from water.

As a forest grows and develops, the density of bird life increases in direct proportion to the increase in the variety of habitats. But a very old forest may be dark, with few openings in the canopy and few shrubs on the ground. Thus it may have fewer niches and fewer birds than a younger one. A mature second-growth forest of the present era probably contains fifteen or twenty more species of birds than its predecessor.

The bluebird, the red-eyed vireo and the indigo bunting are all inhabitants of the forest edge, where they nest in trees and bushes but also can

Nurse Log
On a forested slope in the Great Smokies, birch trees and rhododendrons grow with their roots entwined over a decaying log. In time the old wood will completely rot away, and the new trees will then stand propped on octopuslike roots which have nothing left to grasp.

take advantage of abundant food and water in the meadow. The meadowlark, which nests on the ground, may keep to the forest edge because it likes a high perch (a taste that is satisfied out on the open prairie by telephone lines). The patchwork of different plant communities growing up on abandoned fields throughout the East has multiplied the forest "edge effect"—and with it our bird populations. Joseph Hickey made bird counts and found a difference between two forest patches of identical size in Westchester County, New York. Although strip "A" was dry and situated on top of a ridge, it contained 15 per cent more nesting birds than strip "B", which bordered a resevoir. The reason? The dry, rocky forest had more clearings, and therefore more "edge."

In the lush Smokies, mild temperatures and heavy precipitation have made the forests so dense

that mist rises on warm days, often obscuring the mountain-tops. Many of the peaks here exceed 6,000 feet (Clingmans Dome, 6,643 feet high, is second only to Mt. Washington in New Hampshire as the loftiest point in the East). The mountains are completely tree-covered. On the heights grow spruce and Fraser fir (this species replacing the balsam fir of the North), and the coves and valleys shelter the largest tracts of deciduous trees in primeval or near-primeval condition to be found in all of North America. The portions of the Great Smoky Mountains National Park that were logged 25 to 50 years ago have recovered so rapidly that three-quarters of a total area of 512,000 acres might qualify for preservation as "wilderness" under the Wilderness Act of September, 1964.

In this magnificent park, Canadian hemlock grows up to nineteen feet in circumference, red spruce to fourteen, cucumber tree (a magnolia) to eighteen, and yellow buckeye to fifteen feet. The greatest of all is a tulip tree with a recorded girth of twenty-four feet. Each valley forest contains a broad sample of the 130 trees recorded for the region, and rare varieties such as yellowwood are commonplace.

In the cove forest at the upper end of the valley still known as Sugarlands, the Big Locust Trail leads through an old maple grove to a wood where moss, stonecrop and the polypody fern abound. Here stands the world's second largest locust tree, 52 inches in diameter. (The record specimen, sixty inches across, is in Kentucky.) The older trees in this cove were growing at the time of the Plymouth Colony's founding. They are surrounded by full-grown trees that were mere shoots a hundred years ago. A decaying log is nurse to a line-up of young birch trees. A larger birch stands up on tall, arching roots on the spot where its nurse log has long since crumbled away. Now in April the high crowns of the trees, not yet leafed out, scarcely shade the trout lilies, the Dutchman's breeches, the violets and wood anemones that line the gently rushing rivulets of melt water from above. At this season Clingmans Dome still has some snow on its crown. The upper coves are still getting snow flurries while the lower campgrounds are soaked in cold, continuous rains.

In June the snow disappears from the heights, and rhododendrons bloom reddish-purple. Animal life now moves up the mountains; bears and bobcats cut trails through the rhododendron slicks, which form natural hedges, and raccoons leave childlike prints along the muddy banks of shallow streams, perhaps while looking for salamanders. Thirty species of salamanders inhabit the Smokies.

No one knows why there are grassy balds on some of the mountain-tops, for none reaches above timberline, and the highest of the peaks —Clingmans and other domes—are completely mantled with fir and spruce. In Cherokee legend, the bald spots were created by the Great Spirit so that the Indians could have mountain-top lookouts. Another legendary explanation is that a giant once came down from the sky and planted his feet on those places. Unscientific as the stories are, they confirm the fact that there were open meadows here before the white settlers arrived. Some may have been cleared by Indians in pre-

historic times, while others may be the result of local weather conditions—ice storms that destroyed the trees, or southwesterly winds that dried out the soil on certain slopes, allowing grass to become established. Under the thick mat of mountain-oat grass, or sedge, a maze of tunnels dug by shrews, mice and voles is shared by all of these animals. Interestingly enough, the red-backed vole, which dominates the alpine heath of Mount Washington's summit, far to the north, has also established itself in the Smokies.

The Great Smoky was the first national park to be sized up for the future—that is, to have a decision made as to whether it should become a recreation area, remain as pure wilderness, or perhaps be both—under the Wilderness Act of 1964. Hearings held at Gatlinburg, Tennessee, and Bryson City, North Carolina, in 1966, developed into a major battle between conservationists and commercial interests. At issue was a plan to set aside part of the park as wilderness, and at the same time build another transmountain highway that would take some of the traffic off Route 441. The conservationists, of course, wanted no part of the road. They pointed out that its construction would temporarily silt up streams, killing fish, and that the new highway would open up the western portion of the park for the construction of camp grounds, lodges, gift shops and all the trappings of civilization. There would clearly be a temptation to link the new road with Route 441, the transmountain highway. Once thrown open for development and unlimited tourist traffic—the park in its present state already has more than 6 million visitors annually—the area would no longer be a wilderness.

Those opposed to the road, including scientists from ten leading universities and representatives of the Great Smoky Mountains Hiking Club, the Sierra Club, and the National Audubon Society, preferred the idea of developing the neighboring national forests for mass recreation. These national forests, they maintained, could serve as buffer zones around the last large *wild* forest in the East, thus preserving it as an ecological entity. The conservationists won, and no highway was built.

Partly as a result of this controversy, the Park Service initiated an environmental education program on a nationwide basis. For the first time, study areas are now being set aside in parks where teachers and students can explore the relationships between plants and animals—and the role of man, himself—in the natural community. Funds for research have been increased several-fold. (In 1962, the natural history research budget of $28,000 was roughly equivalent to the cost of one campground comfort station.) It is still too early to judge whether or not this trend of increasing study and research will continue as the population of the United States—and thus the number of visitors to U.S. parks—keeps growing. The act of 1916 creating the Park Service declared that the Service shall "conserve the scenery and the natural and historical objects and the wildlife therein, and . . . provide for the enjoyment of the same in such a manner and by such means as will leave them unimpaired for the enjoyment of future generations." Each year the task becomes more difficult.

Giant sitka spruce and Western hemlock dominate the rain forest valleys of the Olympic Mountains in Washington. Here northern latitude and mountains rising abruptly 8,000 feet above sea level, in combination with a mild coastal climate (17 to 30 weeks of frost-free weather every year) are responsible for the country's wettest environment. The fog and rainfall on mountain slopes facing the sea plus the melt from huge permanent glaciers add up to continual moisture. It can be seen and felt in the air and on the vegetation, even in the driest season of summer. Moisture speeds the breakup of chemicals in decaying layers of the forest floor, and growth goes on at a fantastic rate at every level, from ferns to tree tops. Bigleaf maple, red alder and black cottonwood grow near streams. Thick sheets of lichens (goat's-beard moss) hang from every branch in the forest understory. (Like the Spanish moss of the Southeast, which is an epiphyte, the goat's beard takes its nourishment from the air and does not parasitize the trees.) Wherever the Douglas fir and the Western red cedar of the Pacific mountain forests intrude, they reach tremendous proportions. One record Douglas fir is 229 feet tall. Green moss upholsters the huge tree trunks and downed logs as well, reflecting an eerie, greenish light.

On the logs, toadstools of different colors and shapes stand in small clusters. Each is the fruit of a fungus that spends most of its life underground. The bundle of fungus strands, or mycelium, decomposes almost anything that bacteria cannot eat—including leaf litter and the trunks and branches of downed trees. Miles of the strands spread through the forest floor simply by branching. With the right stimulus, usually a summer rain, mycelia of two different genetic types unite to form the familiar toadstool. From the tight little ball of a new mycelium springs stalk, cap and gills, perfectly designed to shed spores on the wind. Once ripened, a common toadstool gives off half a million spores a minute, and this continues day after day. The spores germinate best on a well-aged log, after sugar fungi and others have reduced it to the consistency of sawdust. Spreading mycelia cause the toughest part of the wood to crumble away, and then the mycelia themselves serve as food for bacteria.

The soil in the forest is a damp, closely-packed mixture of decaying evergreen needles, maple leaves, and sticks. In it can be found a unique assortment of snails, slugs, millipedes, beetles and spiders—the same crawling creatures that inhabit the rotten wood of the fallen trees.

Because much of the Olympic Peninsula was set aside as a federal reserve in 1897, before any logging was done in the upper valleys of the rain forests, the present Olympic National Park (more than 1,000 square miles in area) is one of the largest samples of primeval America left today, outside of Alaska. The old woods harbor such specialized life as the dusky tree mouse, which always stays in the tops of the trees, and the mountain beaver, a primitive rodent whose niche is the leaf mold and deeper levels of the soil basement. A biologist at the Forest Research Center in Olympia, Washington, recently discovered some unusual ectoparasites on this burrowing animal, including the largest and the smallest fleas known to science.

Olympic National Park is the stronghold of the Roosevelt elk, which roams the coastal strip of forest from northern California up through British

Flood Victims
Downed by the spring floods that result from melting snow at higher altitude, conifers lie tumbled across the rocky bed of an Olympic Mountain stream.

Mushrooms provide a touch of orange among the greenery.   A big slug leaves a slimy trail through ferny vegetation.

The Rain Forest

Under the Sitka spruce of the rain forest in Washington State's Hoh Valley,
the slender vine maple forms an eerie understory, its coiled branches
draped with massive sheets of goats' beard moss. Colorful mushrooms
cluster on logs in the damp atmosphere of the rain forest.
These are the fruits of fungus strands that permeate the wood;
often springing up overnight, they cast millions of spores on the wind.
The mushroom spores, tinier than the seeds of an orchid, are no larger than
some bacteria. After finding a proper seed bed, they mature rapidly—
producing branching mycelia, and eventually, more mushrooms.

A spruce stump bears a clump of young trees on top, out of reach of elk and deer which have left the ground bare.

Columbia. The over-browsing of the elk—and to a lesser extent, of the Columbian black-tail, a race of mule deer—has cleared off the undergrowth in the valleys, leaving these places open and park-like except in a few fenced areas. Despite the impression of natural luxuriance, a study in 1945 revealed that the wildlife at the herb and shrub levels of the Olympic forest is four to ten per cent below that of a normal deciduous forest in North Carolina, Connecticut, or Illinois. One of the note-worthy features of the Hoh River Trail is the trampled forest floor. Reproduction of plant life here clearly depends on the saplings that sprout from stumps or nurse logs well above the reach of the browsing animals.

The Olympic Peninsula is isolated, and perhaps has always had too many ungulates and not enough predators. A rich environment, it probably will be able to survive the over-use of vegetation at vital reproductive levels. On the other hand, the forest might be improved by culling the herds, a practice that is becoming more and more necessary in national parks throughout the world. (Over-protection of certain animals, especially ungulates, is the subject of a Conservation Foundation study by the British ecologist, Fraser Darling, who is looking especially into U.S. problems.)

Growing in a narrow fog belt scarcely 30 miles wide along the California coast, the redwood (*Sequoia sempervirens*) has a special ecology as well as great commercial and sentimental value. The lumberman sees a 300-foot-tall redwood that has lived a thousand years or so as enough board feet of lumber to build several houses. The conservationist sees the same tree as a pillar of one of the last remaining cathedral groves. For the last 40 years, some 27,488 acres of California redwoods have been preserved in three state parks —Prairie Creek, Del Norte Coast and Jedediah Smith—which will form the nucleus of Redwood National Park, approved by Congress in September, 1968. However, it still remains for some agency to come up with the $92-million needed for the purchase of 30,512 acres of prime redwoods owned by Arcata and other lumber companies.

Around the sale of this land, worth $2,000 to $3,000 an acre, has raged one of the bitterest conservation fights in American history. Pro-park forces maintained that a redwood preserve would benefit local people by switching the regional economy from timber to tourism within a decade. But the lumbermen were unmoved. To swing public opinion their way, they launched massive campaigns featuring election-style buttons ("Don't Park Our Jobs") and road signs that proclaimed the slogan of the commercial foresters, "Redwoods Forever." (Management of the forests for sustained yield has in fact only recently begun. There was no re-seeding of redwoods until conservationists pressured the lumber companies into starting this practice less than a decade ago. The second-growth redwoods which have sprouted from stumps and roots will not reach harvestable size until the year 2,000.) The men kept cutting the old trees in the disputed area, and the trucks piled high with redwood logs kept rolling night and day, until Interior Secretary Udall and President Johnson brought a stop to destruction of the groves.

To understand the lumberman's point of view, you must drive over four miles of dirt road near the Gold Bluffs of Prairie Creek, and see the stumps of redwoods cut down by Arcata in 1964. A large sign tells the tourist that this grove of "overmature trees" had to be lumbered to salvage the wood. To the commercial mind, it would appear, there is something unhealthy and unnatural about a tree that can live 2,000 years, bearing the scars of fires and insect borings, and braced against floods by its great arching roots. Even after death, the redwood may sprout—a reason for its scientific name, "ever-living."

The redwoods in the parks comprise a vast storehouse of timber—either standing or fallen and in various stages of returning to the soil. Because the trees are now protected from fire, the forest floor is an almost impenetrable mass of young alders, spruce and ferns. New production of redwood trees is limited to the "edges," the disturbed roadside areas where mineral-rich soils have become exposed.

This wilderness *should* be managed as the precious storehouse it is—not as a wood factory. There is a need for controlled fires to clear out the underbrush and maintain the forest's vitality. (In our national forests and parks the emphasis has been all the other way, following the gospel of Smokey the Bear.) On the whole, of course, there

OVERMATURE TIMBER HARVE[ST]
IN FULL COMPLIANCE WITH CA[LIFORNIA]
FOREST PRACTICE LAWS AND RE[GULATIONS]

HELICOPTER SOWED 21 MILLI[ON]
TO START A NEW FOREST JAN[UARY]
58% Redwood 24% Spruce 18[%]

Devastated Area

A sign on the road to California's Gold Bluffs advises motorists that the huge old redwoods destroyed here were really "overmature timber," and that the lumber company has reseeded this forest area.

can never be the degree of production in the wilderness that is possible in a commercial forest where the land is seeded, weeded, sprayed and tended like an agricultural crop. Yet the professional forester doesn't have all the answers to the problems of forest management; on millions of acres of logged-over Douglas fir forest in Washington and Oregon, animal invasions cause an estimated $15-million damage annually. Rodents eat seeds and seedlings, deer and elk browse on leafy shoots, while bears and porcupines gnaw and claw at the older trees.

Logging in California has created worse problems than these by ruining many watersheds.

When 20 to 30 inches of rain fell in the Eel, Klamath, and Smith River drainage areas in 1955, 1964, and again in 1969, homes were flooded, people were drowned, and valuable old redwoods were destroyed in the parks downstream. As pointed out by Ray Dasmann, former head of the Division of Natural Resources of Humboldt State College at Arcata: "In a region such as northwestern California, it would make sense to take great care of the watersheds. But perhaps nowhere else in America can one see more flagrant abuse of watershed cover." One of the most cogent arguments for the new Redwood National Park is the necessity for preservation of these watersheds.

5
Grasslands and Sagebrush Country

The center of our continent dips like a bowl into lowlands where once there were giant seas. During the Paleozoic Era, 450 million years ago, the inundation of North America was so extensive that fish could have swum directly from the Atlantic to the Pacific Ocean, or from Hudson Bay to the Gulf of Mexico. Today, the ancient seaways are marked by sheets of Upper Ordovician limestone, each containing a distinctive fossil fauna —corals, trilobites, sharks and primitive bony fishes. Late in the Paleozoic Era, the eastern mountains rose, and the shorelines of the interior lagoons shifted hundreds of miles. This was the beginning of the Mesozoic Era, a period dominated by swamp-dwelling reptiles. After many millions of years, the entire western region was uplifted and a new habitat—the grasslands—was created for the grazing mammals of the Cenozoic Era.

Across Colorado and Kansas today spreads a rolling plain, built up by sediments washed down from the Rocky Mountains over the millennia or deposited by glaciers. Where it is built by stream deposition and windblown sand (loess), the land seems level; actually, it slopes gently toward the east for 550 miles. In eastern Kansas the elevation is approximately a mile lower than along the Front Range of Colorado, at the western edge of the plain. Glacial drift may be seen piled up most spectacularly along the path of the Missouri River where it cuts through the Dakotas, in the "sharp ridges," or *Coteau des Missouri.*

During inter-glacial periods, when most of the continent was free of ice, melt water collected in natural basins along the fronts of the retreating glaciers. Lake Agassiz was the biggest of these glacial lakes; in the late Pleistocene Epoch, its waters covered 100,000 square miles of the present Red River Valley of North Dakota, and adjacent areas in Manitoba, Saskatchewan and Ontario. Rich, black soils developed in the bed of this now-extinct lake and in similar basins of the central lowlands. In the drying climate of recent times, tough grasses with spreading mats of roots have occupied these open areas, adapting to the seasonal droughts, the high winds, and the extremes of heat and cold.

Today the grass-covered prairies survive in their natural state only as scattered islands in a sea of cultivated land—most often along railroad and highway rights-of-way and fence rows, or in small acreages kept intact for university study. The former tall-grass country from Ohio to eastern Oklahoma is now given over almost completely to corn or pasture plants. Westward, to the 100th Meridian, the precipitation gradually decreases from about 30 inches of rain and snow to 15 inches a year, adequate for the growth of mixed grasses varying from half a foot to two feet in height. What were once mixed-grass prairies are today nearly all fields of grain—mostly spring wheat in the Dakotas and winter wheat in Kansas and Oklahoma.

Across the wide Missouri, at Pierre, South Dakota, the land changes suddenly; it rises 600 feet to the Missouri River Escarpment. Farm houses and trees disappear, and the rolling hills, covered with short grass, extend monotonously to the far horizon. Here the buffalo grass, grama, and needle grass of the old plains still flourish where the cattle range.

In Wyoming, a broad corridor through the Rockies links the high plains to Far Western grazing lands of semi-desert grass and sagebrush, where only about five to ten inches of precipitation can be expected in a year. The area was not always so dry. The great glaciations of the Pleistocene Epoch brought cloudy, rainy weather to deserts a thousand miles south of the ice fronts. Land basins, surrounded by mountains, were filled with water. When the climate changed, the smaller lakes disappeared. Ancient Lake Bonneville and Lake Lahontan slowly evaporated, leaving widely scattered salt lakes—Great Salt Lake in Utah, Mono Lake in California, and Walker and Pyramid Lakes in Nevada. Lack of drainage in this region produced poor, saline soils and a sparse

growth of grass. Little of the original perennial grass can now be found, except in old cemeteries and other places from which cattle have been fenced out. In place of the native perennials grow crested wheat grass, an imported fodder, and small-leaved desert shrubs which are relatively unpalatable to livestock.

The high plains are no longer the stamping grounds of some 30 million bison, half the former continental population. Gone, too, are the prairie-dog towns that extended for many miles, and the buffalo wolf, as the gray wolf was once called. To see the grass prairies in something like their original richness of animal life you must go to present-day bison ranges—the National Bison Range in Montana, the Wichita Mountains in Oklahoma, and Wind Cave National Park and Custer State Park in the Black Hills of South Dakota.

Wind Cave is a borderland of the high plains. Here rainfall is sufficient for the growth of Black Hills pines. The spread of these trees out into the grassland is limited largely by fire and by the trampling of the bison. Comparison of photographs taken 30 years ago with those taken recently in the Black Hills area shows an increase in the number of trees since fire protection began. Today it is impossible to tell what the original vegetation on the plains might have been, or how it might have grown back after fires caused by

lightning or set by Indians. On the managed range, however, a mixture of Western wheat grass and blue grama predominates. There is only a small sampling of buffalo grass. In 1876, Colonel Richard Irving Dodge described this short grass as "covering the ground very thickly" on the vast buffalo ranges of South Dakota "to the exclusion of the other grasses and even flowers . . . forming a thick, close mat of beautiful sward."

Except for small, privately owned herds, no bison now roam the short-grass plains. This land is so valuable as cattle range that no acreage has ever been set aside as a national or state bison preserve. The bison's present home is the mixed-grass prairie. As in the past, most of the grasses here are perennials, nutritious when eaten either as tender spring shoots or as golden dried grass in the fall.

Like the forest, the grassland is layered above and below ground by different species of plant life. Wildflowers appearing among the grasses are not dominant just because they are taller. The real struggle for dominance goes on below ground, where the greater absorbing surfaces of the grass roots capture the available water and make it difficult for other plants to compete for growing room. Once established in a space, however, a wildflower (or weed, as the goldenrod, the false mallow, and the psoralea are sometimes classified) can survive in sod because its single fleshy root plumbs the water supply below the grass roots. The many-flowered psoralea sends its root down nine feet.

On drier soils the thick roots of the shorter grasses, only four to six inches tall, reach down as far as five feet, excluding the taller grasses, which have deeper but less efficient root systems. The little bluestem, for example, grows about two feet tall and sends its roots down as much as seven feet. But it cannot move onto the short-grass plains unless there has been heavy rainfall.

On the plains there is no such thing as "average" rainfall. There may be seven years of rain six inches above normal and then seven years of rain six inches below normal. Alternating wet and dry periods such as this have been the weather pattern for a very long time, as evidenced by the layers of dust between Indian villages dug up by archeologists, and by the accounts of blinding dust storms encountered during expeditions that preceded the settlers with their cattle and their disc plows.

The summer of 1965 proved to be one of the wettest in memory at Wind Cave National Park. Near-nightly thundershowers kept the grass green in mid-July when ordinarily it would have been brown and dormant. Taller grasses formed an "upper story" over much of the short-grass range, and wildflowers bloomed that had not been seen in the park for fifteen or twenty years. The Black Hills and much of the West are still in a wet cycle. But in a few years the trend will almost certainly be reversed. As the drought progresses, the wildflower population will steadily decrease, the taller grasses will die back, and finally between 80 and 90 per cent of the ground will be bare of short grasses. There will be no wildflowers at all except those with eight- to twenty-foot roots.

But curled leaves, dormant roots and seeds will wait out the dust storms. The basis of the grassland's richness is its power of quick recuperation. There is little leaching of minerals, and the soil bacteria benefit from an alkaline environment to survive until the next growing season. Certain plants, such as the clovers and the prairie bluebonnet, utilize nitrogen from the air by way of symbiotic bacteria that invade their roots and form nodules. When the rains come, the tiny soil organisms rapidly release large quantities of nitrogen. As a consequence, the range grasses of high protein content, or hardness, quickly grow to maturity, producing large seed crops.

To live on the open plains, an animal has to be either fast or strong, or it must be able to burrow into the ground. Innumerable small rodents are burrowers, among them the pocket gopher. More than 70 cubic feet of earth have been moved by a single gopher in less than a year. In working its way through the ground, it uses its powerful upper incisors like a pick to loosen the earth, while long-clawed forefeet press the loosened earth back under its body. The teeth are both shovels and choppers of roots—which helps to explain why they are so much bigger than a ground squirrel's.

Having few enemies above ground, the gopher has a degenerate eye. By contrast, the diurnal ground squirrels have large eyes; these animals feed on seeds and on insects such as grasshoppers out in the open, and must therefore be constantly alert to the approach of a predator. It has recently

Good Teeth
Giant incisors are more important to the pocket gopher than keen eyesight. This animal spends most of its time underground, digging tunnels and nibbling on roots. Only for a few weeks in the spring does the solitary male come up to look for a mate, and at this time it becomes highly vulnerable to hawks and other predators.

Good Eyes
No creature is more dependent upon efficient eyesight for survival than the thirteen-lined ground squirrel, a sociable animal that stays with its family and feeds out in the open all summer long. Alert to the ever-present dangers of the prairie, its alarm cry serves as a warning to other squirrels.

On a Western plain, a hungry badger looking for a meal scents a prairie dog and turns toward its prey.

been discovered in the laboratory that a ground squirrel's eye is all cones, the visual receptors that make for keen daylight vision. Most rodents are color blind, but ground squirrels can distinguish one color—blue—from others of equal intensity. And they have special equipment at the optic-nerve level that detects horizontal movement, such as a running coyote. Only one other mammal, the rabbit, is known to have this adaptation.

Since numerous pairs of eyes are better than one pair, ground squirrel colonies have become social groups, and the animals have developed a "language" to communicate what they see to one another. The pitch, loudness and inflection of their chirps and the attitudes of their bodies convey specific warnings about different kinds of danger: snake, dog, man, hawk, coyote, badger. The closer the danger comes, the more urgent are the signals passed from family to family as they take to their holes.

The most sociable of all the ground squirrels is the prairie dog. During the spring breeding season a prairie dog town is divided into territories in which adults communicate with one another and with their young. But individual prairie dogs are unwelcome in any district other than their own.

Occasionally an animal is seen to jump up on its hind legs with its front paws raised high over its head and give a loud "yip," a territorial warning to another dog not to trespass within its precinct. This call, which got the prairie dog its canine name, is unlike the animal's danger signals—the tail flicking and high-pitched calls of alarm that send its neighbors scurrying into their burrows.

Predators, in their turn, have developed tactics to circumvent the organization of a prairie-dog town. The badger applies muscle and claw to the task of digging out its prey. The black-footed ferret and the rattlesnake pursue prairie dogs down into their burrows. But the snake may be foiled by the cooperative action of the prairie dogs; the animals will rush to an invaded hole and try to suffocate their common enemy by packing the hole full of earth. Other enemies—eagles, hawks and coyotes—must be fast enough, or clever enough, to catch prairie dogs above ground. The coyote may follow a badger and nab any member of a dog family that manages to run out past the digger. Another maneuver involves cooperation between two coyotes. One walks through the town, attracting the attention of all the dogs, while the other sneaks up and makes a kill.

Death on the Prairie

From late afternoon through the evening hours, a badger (left) will prowl through a prairie dog town. This predator is a special menace because it can dig down ten or fifteen feet to make a kill. Seeing a badger, any prairie dog will commence barking loudly and flicking its tail (right), a warning signal to the entire dog community. When the badger gets near, all the prairie dogs scramble swiftly into their holes. But at least one unlucky animal will be dug out to make a meal for the badger.

The prairie dog barks in excitement and prepares to dive into its burrow.

After digging out the prairie dog, the badger makes a quick kill and departs with the dead animal in its powerful jaws.

Prairie Rattlesnake
The rattlesnake, which often lives in a deserted burrow in a prairie dog town, can use its natural camouflage to sneak up on young dogs at play. When a prairie dog sees a snake like this one, coiled and ready to strike, it instinctively freezes and then slowly backs into its hole—instead of running, as it would from a badger.

The most elusive of all the prairie dog's enemies, the black-footed ferret, was scarce even before the West was explored and settled by white men, and only a few hundred have been seen since its first recorded discovery in 1851. Today the known population of black-footed ferrets centers around several ranches in the Badlands of South Dakota, where a number have been sighted in recent years and where one is occasionally found dead on the road. Because of its scarcity and secretive, nocturnal habits, practically nothing was known about the species until federal and state biologists became worried about its survival. Now an almost year-round vigil is kept. Observers using spotlights on the plains at night have made some interesting discoveries about these animals.

In each prairie-dog town there is one family of ferrets. They live in deserted burrows and move about as they decimate the population of surrounding burrows. The dogs try to counter by sealing up any hole that has a weasel scent, and they will gang up on any ferret unfortunate enough to be caught abroad in daylight. As soon as the young ferrets are grown, the older generation drives them away, or else they drift off on their own to colonize another town. The constant need for new places to conquer must necessarily limit the numbers of black-footed ferrets, and they will become increasingly rare as live prairie-dog towns become fewer and fewer.

Wherever prairie dogs are protected, their towns expand. After the breeding season, they move from overcrowded to less populated districts. Some pioneers dig new holes in the suburbs to form new districts. Some, still more enterprising, move to the next valley and start a new town. At Wind Cave there are eight established towns, and new ones are being started in the Highland Creek drainage, the Dry Creek drainage, and the meadow west of Rankin Ridge.

Across the unprotected prairies, deserted prairie-dog towns are colonized by other animals that

Burrowing Owl

A burrowing owl emerges from its subterranean home. Though capable of digging a hole,
the bird prefers to use a deserted burrow in a prairie dog town. It benefits
the prairie dogs by giving a screeching alarm call when a common enemy appears.

traditionally use the holes: burrowing owls, which prefer not to dig their own holes, and snakes, which cannot dig.

The high mound that the prairie dog throws up around his entrance is more than a perch from which he surveys his world. It helps to prevent flooding, a persistent threat on the flatlands. The older the burrow, the larger the mound becomes, until some are a foot high and as much as eight feet in diameter.

Unlike the smaller rodents that burrow among the roots of the grasses and are probably useful in aerating the soil, the prairie dog digs down fourteen feet. He packs firmly into his mound anything he brings up from these depths, and all through the summer the mound grows. As the grasses flower and seed, they are eaten away in an ever-widening circle around the burrow. The prairie dog cuts down the milkweed when it blooms; allowed to lie on the ground, all of the milkweed seeds ripen at once. The heads and leaves of grama grass are eaten in June; by July the leaves become too old and tough. In August there is cured grass of various kinds to feed on, and there are cockleburs, shelled for the seeds. The prairie dog's food preferences are evident from the islands of surviving vegetation. At the end of the season the town may look barren, the soil hard and cracked.

The fact that two-thirds of the prairie dog's food is grass and that colonies of dogs often occupy the most fertile river bottoms is enough to condemn these animals in the eyes of cattlemen. When the first settlers came to the plains, they instituted the Sunday afternoon sport of "varmint hunting." In the 1930's, poisoning campaigns began that killed out the big prairie-dog towns from Kansas to Texas and as far north as the Pine Ridge and Rosebud Indian Reservations of South Dakota. Government teams systematically rode their horses the length and breadth of the towns for weeks at a time, dropping strychnine

Normal and Black

*Over thousands of years, the rodents of the
open grasslands have evolved fur colors
that blend with the earth and drying grasses.
Any deviates that appear, through genetic
mutation, may or may not be able to survive.
The normal, thirteen-lined ground squirrel
(above) outnumbers the melanistic individual
(below) by perhaps ten million to one.
Having an excess of the black pigment, melanin,
the black squirrel is stronger and more
active than the normal one. After some 25
generations of cross-breeding, a dark phase
might become established on the plains. But
predators more likely would prevent the
evolution of a local color phase by killing off
most of the conspicuous black squirrels.*

pellets into the burrows.

The sociable habits of the prairie dogs, which
had given them an advantage on the natural
plains, made them particularly vulnerable to man.
Their numbers greatly reduced by the federal
poisoning campaigns, they have survived over
the last fifty years largely on game ranges and in
national parks. Now it is a novelty to drive through
Wind Cave Park and see prairie-dog mounds
stretching as far as the eye can see, with their
owners standing up and barking excitedly. If you
happen to find a colony that is not in a park,
there is no sign of the animals except for the tell-
tale fresh burrowings. Self-concealment stems
from a learned caution; though the federal government no longer sponsors the poisoning of prairie
dogs, the states have continued the practice.

In some places, prairie dogs have even given
up the habit of building mounds. South of Fort
Carson, Colorado, a prairie-dog town on a target
range was completely leveled by tanks recently.
Two years later the fields were left to grow back,

and the prairie dogs returned. But not one made
a mound around its hole. From the county road
nearby, the town was practically invisible, and
few people knew it existed.

In nature, to be seen is very often to be dead.
The prey tries to be inconspicuous and the predator sharpens its senses over the ages to perceive
very small differences in coloration or shading.
Even animals that are believed to be color blind
may have keen perception of tonal values. The
experiments of L. R. Dice prove that owls can
detect a mouse whose color differs only slightly
from a gray-brown background (simulating conditions at night in the fields and woods). Such
perception by predators tends to make populations
of rodents all of a color, the one least conspicuous
in any geographical area. Very often color is the
only basis on which a species is separated into
local races by the taxonomists.

Because of vulnerability, it is understandable
that color variations, or phases, are rare in rodents.
The most common color is a blend of many colors,

called agouti, from which black, gray, blue, red, brown, and white can be bred in the laboratory.

By contrast, the predators do develop color phases. Examples are the black and the normal phases of the red-tailed hawk and other related hawks in the Western United States, where the black birds are as numerous as the others. The arctic fox also has two distinct forms—the "normal," which is brown in summer and white in winter, and the "blue," which retains a dark bluish-gray coat color all year. These phases are genetically distinct, the "blue" apparently dominant and the "normal" recessive.

Often important to its survival is the fact that an animal with a red coat is seen by another which is color blind as though it were green—both colors appearing gray. In the days of the buffalo wolf, which could not distinguish red, the tawny coats of young bison blended into the green of the prairie in spring and helped them to stay alive out in the open. The red fox cannot be seen by its color-blind prey, the rabbit and the rodent, if it doesn't betray its presence by *movement*. Given a coloring that is inconspicuous, freezing is the best camouflage for a hunter or a hunted animal.

Among the large grazing animals, a flash of contrasting colors or a white tail lifted while running communicates a warning to others in the herd. The most highly developed signal of this kind on the plains is the white rump patch of the pronghorn antelope. When the animal is alarmed, guard hairs rise in a fan that is visible at great distances.

The pronghorn is a uniquely American animal, no relative of the African antelopes. Both males and females bear horns, and both have over the forehead the horn extension, or prong, for which the species is named. (Often the doe's horns are smaller than her mate's, and she never has the prominent black mask and neck patch that distinguishes the pronghorn buck.) The horns are shed each fall, leaving bony cores. From the tip of each core grows a new horn sheath that is complete and hardened by the next July.

Large eyes, keen vision, and an efficient signaling system help the pronghorn herds escape their enemies on the open plains. But speed is their chief means of defense. For short sprints they can reach 40 miles an hour. Thus it is only at kidding time that predators—formerly the wolf, and now the bobcat and the coyote—can make a substantial number of kills; these are mostly newborn kids.

A pronghorn just six days old runs smoothly. After a week or ten days it has the white rump patch of the parents. During the first months of life it learns the tactics of flight by trying to outrun its mother, and it will continue practicing until, feeling the need for rest or safety, it drops down and hides in the grass.

When man came on the scene with guns, the confidence of the pronghorn in its ability to get away from enemies was its undoing. Hunters found that they could wave a handkerchief and keep the interest of an animal until they had crept up within shooting range. In 1924 the U.S. Biological Survey reported the number of pronghorns down to 30,000. Most of the survivors were on the west side of the Rockies, in Nevada, Oregon and California, and the biggest herds are there today. With protection, the number of pronghorns has grown to 350,000.

At Wind Cave the pronghorn and the bison are kept at about equal numbers, as they were on the natural plains. In 1965, park rangers counted 254 bison and 242 pronghorn—after 220 selected bison had been culled from the herd and 75 pronghorn had been driven over the border to neighboring Custer State Park, which has a much larger range. Today Wind Cave's bison herd and Custer's (1,500) together comprise the largest existing assemblage of these magnificent and powerful beasts.

Follow the Custer herd over the rolling landscape in a four-wheel-drive vehicle and you will soon realize how fast the seemingly ponderous animals can move. Just walking and clipping the grass at a leisurely pace of four or five miles an

Warning Signal
A pronghorn buck, crossing the plains of Wind Cave National Park in South Dakota, flares out its white rump patch as a signal to the herd of danger observed.

A Lone Hunter

*Following game trails long ago deserted
by wolves, the coyote still fills its ancient
role on the plains. It preys on small game
and waits for opportunities to pick off
the young, the old, and the injured of the big
game animals. Although hunted by man almost
universally, trapped, poisoned, and gunned
down from airplanes, this species has
persevered in North America, seemingly
tougher and craftier as the years go by.*

An Amiable Society

*At the beginning of summer, bison
graze peacefully on a South Dakota plain,
the animals divided into small family
groups. Each calf is protected by its father
and its mother—and frequently by the
entire herd. Weaning takes place after one
year, and youngsters stay with their mothers
until three years old. Then the young
bulls will join the battles of the rutting
season and find mates for themselves.*

The Wallow

Hooves kicking in air, an old bull rolls over in a vigorous wallow at the start of the rutting season. The dust bath helps get rid of matted clumps of old fur and biting insects. At favorite wallowing places, the bison make deep hollows that fill after a rain, to become water holes on the plains.

hour, they will disappear over a rise and you will lose them time after time. But a real eye-opener is to see a big bull break into a run. July is the beginning of rutting for the bulls. They challenge each other and engage in mock fights around the water hole as preliminaries to more serious encounters later. With the buffalo wolf gone, the bison's sharp curved horns are used only to threaten other bulls and wrestle them in the ritual of the mating season.

The bison is anything but clumsy. Its muscular strength is in the shoulders and its enormous weight pivots on the front legs. Like an African rhino, this dreadnaught can turn on a dime and sometimes charges without warning.

No animal had more effect on the ecology of the plains than the bison. By the weight of its hooves it kept tall grass out of the Kansas borderland and made the land clear for prairie dogs. The durable short grasses on which the bison fed needed the pounding of its feet for the germination of their seeds. (Today buffalo-grass seed has to be mechanically abraded before it is sown on the managed range.) Cattle cannot serve the same function as the huge bison herds, which migrated hundreds of miles northward in the early summer and south again in the winter. Always on the move, the herds allowed the land to recover during their absence.

The pronghorn drifted to areas of less snow in

A Time of Trial
In one of the early battles of the rutting season, young bulls square off (above) and then lock horns in a test of strength. There is more bluff than blood involved in a struggle of this sort for male dominance;
the weaker of the two combatants nearly always retreats before being seriously injured by the other's horns.

winter but did not migrate. Instead, this animal switched from a diet of grass and wildflowers to other foods, such as woody shrubs.

In their relationship to the land, the bison could be compared to the wildebeest, and the pronghorn to the impala of the African plains. Millions of years ago, North America's plains were savannas with the shrub growth and trees and the many kinds of animals with diversified feeding habits that still exist in Africa today.

During the 19th Century, the bison was very nearly hunted to extinction. (The art of the buffalo hunter consisted in shooting the animals down one by one, dropping each animal where it stood, so that there was no fast clatter of hooves to stampede the others. It was a technique learned from the plains Indians, who hunted buffalo in the same way, but less efficiently, with their bows and arrows.) By 1889 there were only 541 buffalo left. The concern of William T. Hornaday of the National Museum and others led to the establishment of the American Bison Society in 1905, to

The Black Hills

The Black Hills of South Dakota were named for their dark covering of pine trees. The hills comprise a great dome, 125 miles long and 65 miles wide, built by the same disturbances that lifted the Rocky Mountains some 200 miles to the west. Peaks on the eastern side of the dome drop off to flatlands at 3,000 to 3,500 foot altitude. Today National Park Service fire towers dot the ridges. In this view, from a point just below the Rankin Ridge tower, dark pines seem to be marching onto the plains.

promote the breeding of bison and their preservation in parks and refuges.

Today there are about 11,000 bison. Some, as in Custer State Park, are raised for meat. They are so prolific that cropping is necessary every year to keep the herds from outgrowing the available land—and also from growing unmanageable. The larger the herd, the stronger the instinct to crash the strong buffalo fences and take to the old migratory trails.

Grazers and browsers, the wapiti (244 in Wind Cave Park) occupy still another niche in the grassland—the forest edge. Grasses make up the larger part of their diet. One study, on the Yellowstone Plateau of Wyoming, showed a proportion of 95 per cent grass consumed to five per cent wildflowers and shrubs. These wapiti are smaller and lighter in color, and are wilder than the Roosevelt elk on the Pacific Coast. Individuals sometimes mingle with deer and pronghorns, but they avoid the bison. Usually they hold themselves aloof in well-organized bands. Disturbed by a scent, sound

Portrait of a Young Bull
A two-year-old American elk, or wapiti, shows a small set of antlers. Up to the age of five, the rack will increase in size, branching to as many as six "points."

or distant movement, the golden-red wapiti quickly vanish among the ponderosa pines. (At top speed, the wapiti travels at about 35 miles per hour, faster than any other animal on the plains except the pronghorn.) Almost invariably, a cow assumes leadership and takes the others to safety.

Beyond the foothills, in the northern Rockies, wapiti spend the summers in high meadows, and in the fall migrate down to the lower valleys, a distance of ten to sixty miles. Before the settlement of the West, they had no need to travel so far. Their present winter range, at 5,000 feet, is so high that many yearlings, unable to stand the cold and the lack of food, perish in the snow.

Evidence that the advent of man has caused the wapiti's change in habits may be found in the Lewis and Clark Journals. While they were on the plains and right up to the time they reached the mountains, the explorers were able to supply

themselves with plenty of fresh meat. But after they began their ascent, game was practically non-existent. When they met the Shoshoni Indians on the headwaters of the Jefferson River, even these natives had only salmon and berry cakes to trade. For meat, the explorers had to depend on horse or dog flesh—when the Indians were willing to supply it. Traveler after early traveler spoke of the abundance of buffalo on the plains in comparison to the scarcity of the animals in the mountains. Only with the tide of hunters and settlers sweeping west of the Missouri did the big game of the plains retreat up the forest-covered slopes.

It is on the high summer range that the bull wapiti bugles and collects his harem early in September, a full month before migration. Every evening the clatter of antlers accompanies the wapiti battles, which are tame affairs compared with the clashes between bull bison. The loser usually runs away at top speed, unhurt, and kicking to show his displeasure. At the close of the rutting season, storms gather over the mountains, and small bands of cows, calves and young bulls assemble in herds of fifty to four hundred or more for the journey down to the valleys. Last to leave are the old bulls. All the way down, the migration is leisurely, the herds pausing longer in certain places than others, according to the available food supply. The temperature may be 41 degrees below zero and the snow two feet deep, but they will stay in a place for some time if the drifts are not crusted over. Like deer, wapiti easily paw through the snow and make a "yard" for resting. In three or four weeks, the migrating animals reach their destination. On the lower mountainside, they will eat practically any kind of woody plant that can be chewed—sagebrush, service berry, aspen, or Douglas fir. Here the calves will be born the following May, before the long trek back to sub-alpine meadows.

On the southern Yellowstone, thousands of elk in the traditional wintering area at Jackson Hole have become partially or wholly dependent on artificial feeding. This is not necessarily a good

Ranks of Wapiti
In a Dakota valley, wapiti stand as if in military formation, alert and ready to run for cover. Three days after birth, a wapiti calf can follow the herd closely.

thing. Biologists have long known that the animal which scratches around for itself in the winter may be in better condition in the spring than one which eats hay, alfalfa, and other apparently nutritious foods. But they did not know the reason until recently.

The answer came from a study of the microbes in the digestive tracts of deer, which—like wapiti—are ruminants and browse on woody plants in the winter. During this season, the microbes must work harder and harder to break down the fibers in what the deer eats. And because digestion becomes increasingly difficult, the animal may begin to suffer from malnutrition. At this time it will readily switch to livestock feed, too rich for its digestive microbes, which are specifically adapted to woody plants. When the supply of the deer's natural microbes is gone, it can no longer digest *any* food, and the animal dies of starvation with its stomach full. To make matters worse, good livestock feeds are high in ammonia, and because they are being eaten at a time of year when bacteria that depend on ammonia are few, the toxic chemical may find its way into the blood stream, poisoning the deer. In either case, it is being fed to death.

Where the mile-high plains of Colorado meet the Front Range, a complex web of food chains begins in the foothills and canyons and radiates out over the grasslands. The lives of the badger, the kit fox and the prairie falcon on the plains are linked with the Colorado chipmunk, the mountain lion and the golden eagle in the cool canyons of ponderosa pine. In the spring, the snow melt from above is swelled by steady rains, and mountain freshets roar over dry creekbeds below, carrying rock fragments and plant debris. (These have built the prairie soil over the ages.) The water table rises, and earthworms—a key food in this transition zone—come to the surface. On the first sunny days, the energy concentrated in the worms is passed along both to small predators like the ants and ground beetles, and to larger ones like the magpies. Other predators are out, too, looking for the worm feeders, or for those that feed on the feeders. Who eats whom depends on the weather, the diurnal or nocturnal habits of animals, their methods of hunting—and sometimes, on pure chance.

A worm is attacked by red ants on a Colorado canyon floor

The ants are eaten by a spiny lizard, immune to their bite.

A garter snake, lurking in the grass, strikes the lizard.

The Food Chain: 1

Along the Front Range of Colorado, the earthworm is a basic food of many animals.
When spring rains bring the worms to the surface, birds scratch for them
and ants attack them. Other predators are out, too, searching for the worm eaters.
All life in the natural world depends upon the maintenance and continuance
of food chains. The activities of one creature touch upon the activities of another
creature, often in a relationship of predator and prey. Life (and death)
experiences are interacting; one food chain cuts across another. The result is
something like a gigantic web, the strands of which are all the various
food chains crisscrossing. On this page is one chain—not in its entirety, but
several of the links that can be shown photographically. On the next four
pages can be seen additional food chains. Each is a strand in the web of life.

The garter snake moves out onto the open prairie, where it is caught and devoured by a badger.

A magpie is lucky enough to find a worm.

The Food Chain: 2
*At an early point in a food chain, a
magpie just out of its nest in a canyon
finds and swallows a worm (above). Later,
when the bird is old enough to fly across
the plains, its conspicuous black-and-white
markings attract the attention of a
prairie falcon, which makes a swift kill.*

Luck runs out when it falls prey to a falcon.

A ground beetle cuts a worm into bite-sized pieces.

A Colorado chipmunk enjoys the crunchy taste of the beetle.

The Food Chain: 3
*This food chain begins in a canyon at night, when a ground beetle
seizes a worm and shreds it with sharp mandibles. The next
morning, a Colorado chipmunk discovers and eats the beetle, relishing
such a delicacy as a change from its usual spring diet of seeds
and berries. At sundown, the chipmunk usually goes to sleep in a rock
crevice, but on this day it is stalked and killed by a kit fox.*

A kit fox, one of the keenest hunters of the plains, enters the canyon and is rewarded when it captures the chipmunk.

East of the Continental Divide, the slopes above the plains sweep sharply upward, encompassing all life zones from plains grasses to tundra grasses within a mere mile and a half. Here a rainstorm far back in the mountains may send flood torrents down without warning. Established drainage patterns disappear in minutes under walls of water that plunge over sheer rock, creating new patterns. Then the flood dissipates out over the plains.

Where streams and rivers have cut into the plains, exposing the whitish ash beds and the red, pink, tan, brown, bluish-gray and gray layers of clays, shales and sandstone, the land is sterile, caught in a continuous cycle of erosion. The region beyond the towering cliffs known as "The Wall" in South Dakota has been completely washed out by the White River, except for jutting peninsulas, ornate pinnacles and spires of bentonite, the predominant clay.

These are the Big Badlands, the most spectacular scar on the surface of the high plains. They expose in the bentonite the bones of oreodonts, camels, and other extinct animals that lived in the Dakota marshes 40 million years ago. And in the low mounds of shale are the shells of marine creatures that inhabited the ancient ocean bottoms.

On the edge of the Wall, certain clumps of grass resist erosion for decades, but some soil washes down and plant life becomes established below. Over hundreds and thousands of years, the action of the wind and rain will undercut and gully the cliffs and level off the plains, which will again be covered with grass.

The Flood
Rains falling far back in the interior of the Rockies swell a mountain stream (*left*) and flood a creek bed in the Front Range (*below*) that has been dry for 20 years. Bridges and roads were destroyed in this flash flood.

A Moonscape on Earth

The Indians and early explorers of the West both avoided the great
scar on the prairies which they called the "bad lands" because
it was so difficult to cross. Infrequent but torrential rains cut deep
gullies in the soft, claylike bentonite which makes up this weird
landscape, set aside since 1939 as the Badlands National Monument.
Below towering pinnacles carved by erosion, only scattered plants
like the lone prairie sunflower (opposite page) can survive here.

Until about 1825, bison roamed in small bands throughout the grass and sagebrush country west of the Rockies. Then the longhorn cattle came, the range was heavily stocked with both cattle and sheep, and the bison disappeared everywhere except in the easternmost extensions of this semi-desert environment—in Jackson Hole, Wyoming, and on the National Bison Range of Montana.

By late August, the antelope refuges of northern Nevada and the Oregon border dry up, and wildlife concentrates around the few remaining "lakes"—which are playas with dry bottoms covered with grass and weedy annuals—or around the muddy waterholes. Since the refuges were established in 1931 (at Sheldon) and 1936 (at Hart Mountain), water supplies have been much improved by damming springs, but the effect of the dry season is still evident. The largest playa, Swan Lake, is filled with hundreds of pronghorns, and occasionally a pair of coyotes may be seen trying to cut off a yearling from the herd. Here death is out in the open. Turkey vultures feed on the carcass of a coyote. Not far away, golden eagles rise startled from the body of a pronghorn mother, and her kid wanders off in search of a new mother.

Occasionally one catches a glimpse of the extremely shy herds of wild mustangs, among the last in the West. More in evidence are thousands of cattle—theoretically not enough to compete with the wildlife—which are allowed to graze here, as on other refuges, under the provisions of the Taylor Grazing Act of 1924. Windmills of deserted homesteads, like Last Chance Ranch, built in 1915, are still turning and filling tubs of water.

Perhaps all of the herbivorous animals in this region once made more use of perennial grasses:

The Water Hole
*Sage grouse gather at a water hole in the Sheldon
National Antelope Refuge. A nearby spring was tapped
to insure a summer drinking supply for wildlife here.*

sandberg bluegrass, Idaho fescue, various needle-
grasses and wheatgrasses. But the heavy invasion
of cattle early in the century opened up the land
for a spreading plague of the gray-green sagebrush
that no Act of Congress could control. Thus both
wildlife and livestock have become increasingly
dependent on sage.

The sage grouse courts and nests under the big
sage brush. It feeds on the buds and leaves of the
low sage. This pungent fare—black, silver, and
other kinds of sage—makes up three quarters of
its diet. The pronghorn antelope prefers the short-
sagebrush range for visibility and quick getaway.
More than half of its diet is sage, mostly black
sagebrush. (The pronghorn also feeds on the wildflowers of spring and summer, the bitterbush of
July and August, and the grasses.)

The Great Basin is not one basin, but many,
filled with debris washed down from mountains

Sagebrush Country
*Tufted fields of dusty-green sagebrush sweep to the
tops of high mesas along the border between Nevada and
Oregon, still a primitive part of the American West.*

201

over long periods of geologic time. Driving through Nevada today, one finds it difficult to realize that heavy snowfall and glacial meltwater made this dry, desolate country what it is. In the undrained basins, lakes evaporated and left alkaline flats in which the salts from the mountainsides had become highly concentrated, fit only for certain alkaline-tolerant members of the beet family, such as greasewood and shadscale—or remaining nothing but barren white flats.

What can live here? In the tall greasewood there are plenty of coyotes, badgers and jack rabbits, but few cattle can be found, and no antelopes or deer. In the low shadscale there are few animals other than the hardiest rodents and the desert kit fox. The kangaroo rats raise up huge mounds, peppered with entrances. Shrubs grow on the mounds, forming islands twice as tall as the surrounding vegetation. By breaking up the hardpan, the kangaroo rats allow the deep roots of the shrubs to reach the water table, and then the rats keep them well nourished with soluble minerals from the decay of stored plant material and droppings.

Clues to the past are terraced and polished rocks, indicating to the geologist depths of 500 to 1,000 feet in the prehistoric lakes: Bonneville in Utah and Lahontan, which covered much of northern and western Nevada and small areas of adjoining Oregon and California. In the Snake Range of eastern Nevada, relict stands of Douglas fir and aspen grow on north-facing slopes, and four species of fishes native to the Bonneville system swim in now-isolated mountain streams. It seems almost certain that dense boreal forests and deep lakes existed here within the time of man. Among the New Mexico Indians there are legends of emigrating to their present homes from the north by canoe through interminable damp caves, perhaps a poetic interpretation of the wet climate that prevailed in this region thousands of years ago.

In the now-arid landscape, corings from trees in mountain "islands in the sky" help us to trace weather patterns and date archeological finds. Crowning the peaks of a few Great Basin ranges that over-reach the others are the oldest living things on earth, the bristlecone pines. The most ancient, gnarled and weathered of these trees— dating back over 4,000 years—have only a small amount of living tissue left, usually a single panel of bark extending from the base to the top. Still the trees can and do produce cones and viable seeds in years of good rainfall. Until then they remain semi-dormant, showing in their rings only a scant record of growth, a dendrochronologist's diary of drought.

On the northeast face of Wheeler Peak in eastern Nevada there stood, until recently, a bristlecone estimated to be 4,900 years old—300 years older than the tree discovered in the White Mountains of California in 1958 which was then thought to be the oldest alive. In the same forest, dead bristlecones still standing are estimated to have begun life more than 5,000 years ago. As core sampling goes on, we may find in these patriarchs, dead and alive, a continuous record of the climate that reaches back more than 10,000 years into the Pleistocene Epoch.

One of the relatively little-known bristlecone pine forests is a 15,000-acre tract in the Spring Mountains of Nevada, where trees 3,500 years old have been found. The only access to the tract is Lees Canyon Road, which starts among Joshua trees and creosote bush—in a corner of the Mojave Desert that spills over from California—and ends among ponderosa pine. After that, there is only a foot or horse trail that leads, in tortuous switchbacks, to the first bristlecones at about 7,500 feet. Here, and in sheltered coves at higher elevations, this pine tree grows tall and straight. The tips of its branches are tightly packed with bundles of needles that resemble nothing so much as dark green bottle brushes. Not until you pass through these dense young forests, only a few hundred years old, and are close to the mountain tops can you see the ancient, twisted bristlecones. There they sit, hugging the cliffs and ridges, their trunks arched into the wind. Weathering has polished the bark to a dull, metallic glow, and needles cling to only two or three of the branches. Below exposed roots, as large as the trunks, the debris of a Devonian coral bed is scattered over the barren, rocky soil—a reminder that these peaks, now 10,000 to 11,500 feet high, were once in the depths of the sea.

The Patriarch
High in the Spring Mountains of Nevada, an ancient bristlecone pine is weathered but still alive after thousands of years of infinitesimally slow growth.

Pyramid Lake, 33 miles northeast of Reno, is one of the last desert sinks in the area covered by old Lake Lahontan. In the midst of "nowhere," its deep blue waters are as impressive to the 20th Century traveler as when they were first seen by General John Fremont on January 10, 1844. "It broke upon our eyes like an ocean," he said. But he did not see or describe the thousands of pelicans that nest on Anaho Island. They were wintering in Mexico at the time.

The white pelicans of Anaho are North America's largest nesting colony. Their island rises like a small mountain 600 feet above the present water surface. Its inhabitants include not only white pelicans but also double-crested cormorants, great blue herons, California and ring-billed gulls, and Caspian terns.

As early as 1919, Woodrow Wilson set aside the Anaho Island Reservation, and on July 25, 1940, it became part of the national wildlife refuge system. Remote as the area is, Pyramid Lake quickly became famous for the biggest cutthroat trout in the West. It was soon found necessary to prevent boats from landing on the island during the nesting period because adult pelicans, scared away from young and eggs, would circle overhead for hours, leaving the colony open to attack by predatory gulls.

Meanwhile trouble was brewing for the lake, the fish and the pelicans. The region was locked in a century-long drought, and most of the flow of the Truckee River, whose source was high in the Sierras at Tahoe, was being diverted by Derby Dam to the lush agricultural valley of Fallon and the marshes of Stillwater Refuge in the old Carson Sink. The construction in 1905 of this dam, which brought green grass and trees—and even fields of alfalfa, fruits and vegetables—to Fallon, finished the trout, which could no longer spawn upstream beyond its locks. It also endangered a rare, primitive fish, the cui-cui, which somehow manages to spawn below the sandbars that have grown up where the river enters the lake.

On the east, Winnemucca Lake, the sister of

Pods of Pelicans
Their wings still untried, young pelicans huddle together on the volcanic rocks of Anaho Island, near the eastern shore of Pyramid Lake, Nevada.

Pyramid Lake, has been dry since 1938, a shining salt flat where peregrine falcons once pursued ducks, and shorebirds sheltered among the reeds. In 50 years, Pyramid Lake has itself dropped 75 feet. If the water continues to lower, a land bridge will form on the east shore connecting Anaho Island with the mainland, and coyotes and other predators will quickly destroy the clumsy, flightless young of the pelicans.

During 1967, some relief appeared. There was a record snowfall in the high Sierras, water flowed out of the dry hills, and hundreds of cui-cui tried to reach the upstream spawning grounds. Secretary of the Interior Stewart L. Udall ordered the rusty locks of the old dam opened for a few days, and the lake rose several feet.

In late summer, when the Bureau of Sport Fisheries and Wildlife bands the young, Anaho is unlike any other place in the modern world. As soon as you have climbed the cinders of the volcanic shoreline of this island that survives from another age, thousands of white pelicans rise up, obscuring the sky with the whir of black-tipped wings. They wheel past with a white storm of droppings. Most of them will not settle down again for hours, but their fat offspring are too big now to be harmed by the gulls. The young birds submit with great squawkings and some biting to the indignity of being held upside down and marked with a metal band.

Official counts reveal that the original colony of about 20,000 Anaho pelicans had slipped to about 14,000 (in 14 separate colonies) by 1950; still

further, to 7,500 (in only four colonies), by 1965. No one knows why this has happened. Are gulls killing too many of the young? Sonic booms are scaring the adult pelicans off the nest with increasing frequency, and the gulls are quick to take advantage of their absence. Or has something happened to the pelicans' food supply? Some biologists believe that the white pelicans of Anaho are victims of insecticides—like the brown pelican, the bald eagle, the osprey and other fish-eating birds that have declined in certain areas of the country.

Hydrocarbon insecticides, originally added to lakes and streams in harmless proportions, are passed along aquatic food chains in stronger and stronger concentrations, each living body storing more than the one before it, from plankton to fishes to fish-eaters. The predator at the end of the food chain may be a pelican that consumes a poison 80,000 times as strong as the original solution in the water. Sometimes the result is a kill of several hundred or more birds. A colony of 1,000 pairs of Western grebes is known to have disappeared from Clear Lake, California, in 1950, and insecticides were suspected. Many birds undoubtedly die from a delayed reaction, when their fat reserves—where the poison is concentrated—are utilized under the stress of hunger and fatigue during migration. Another place that the poison may concentrate is in the bird's egg, but much is yet to be learned about egg failures and mortality in the young.

Over the past decade, white pelicans have been among the chief victims of insecticide kills on the

Stilt
This black-necked stilt, whose name is highly descriptive of the bird's appearance, is exploring the muddy shoreline of a reservoir in the Grand Coulee.

Avocet
Closely related to the stilt, the avocet wades the same shallows, feeding from the bottom, but also swims offshore. It has waterproof feathers, like a duck's.

Cut Out of Rock

Colorful lichens encrust cliffs in the Upper Grand Coulee, a great gorge in northeastern Washington that was cut when the Columbia River was diverted this way during the Pleistocene epoch. Long dry, the gorge has had water in it since Grand Coulee Dam was built in 1960, restoring the habitats of ducks and shorebirds (opposite page)

resort lakes and refuges of neighboring California. On Lower Klamath Refuge, which receives drainage from farms, 156 of 307 dead birds that were picked up in 1960 and found to have died of toxaphene and DDT poisoning were white pelicans. In the same year, Big Bear Lake in San Bernardino County was treated with toxaphene to eliminate "rough fish," and a similar kill occurred. A live white pelican which was shot and analyzed had as much toxaphene in its fat (1700 p.p.m.) as the dead grebes of Clear Lake.

Research, through banding and color marking, is under way to find out where the Anaho pelicans migrate and how they may be affected by contaminated food at Fallon, on California refuges, and at feeding places in Mexico.

Not all irrigation threatens the nesting areas of water birds. The drainage ditches of Washington's Yakima Valley have for many years produced more ducks per linear mile (200 broods) than the pothole country of Canada and the Dakotas east of the Rockies. In 1960, the completion of the Coulee Dam backed water down an ancient Ice-Age Channel of the Columbia River, creating in the Grand Coulee (French Canadian for "big canyon") a 30-mile-long reservoir, Banks Lake, which attracts mallards, bluewings, cinnamon teal and other waterfowl. The scummy margins of such newly-flooded lands, buzzing with flies, are prime hunting grounds for avocets and black-necked stilts, which thrive in greater numbers on the open, alkaline marshes of the West than in the tidal marshes of the East.

In the future we can expect more farming and more irrigation in the West, and more herbicide control of sagebrush. In September of 1967, *Farm Technology* reported that in one county of Oregon alone, 27,000 acres had been sprayed with 2,4-D by helicopters and planes in the first stage of a new five-year program. Part of the range was re-seeded, primarily in crested wheat grass, and the rest was left to be reclaimed by native grasses. Some wildlife habitats will be destroyed and others will be created. With planning, perhaps the gains can be made to balance the losses.

Overgrazing must be curbed. A single cow needs only 30 acres of prime grass on the Great Plains, but in the sagebrush country each cow needs 40 to 80 acres of grazing land. Among big-game animals, only the bison now competes with cattle for grass. The pronghorn antelope eats weeds along fence-rows and in gullies and breaks. Mule deer concentrate on shrubbery among the rimrocks. White-tailed deer browse in the mountains, and wapiti along the borders of foothill forests. Still, the damage done by the overgrazing of cattle affects all wild game by preventing the recovery of natural grasses and causing the spread of the least palatable shrubs. Rodents and rabbits build up their populations on the fleshy roots and abundant seeds of a weedy range. When a drought comes, it sometimes appears that these small animals—together with the hordes of grasshoppers—have themselves destroyed the range. The only animal to profit from such a disaster is the coyote, which feeds on the man-caused eruption of rabbits and rodents and also attacks sheep, cattle and starving game.

To eradicate these apparent trouble-makers—the rodents, rabbits and coyotes—more and more deadly poisons have been devised: first strychnine, then tasteless and odorless thallium, and finally—in the late 1940's—"1080" (sodium fluoroacetate).

Theoretically "1080" was selective, killing only the rodents and coyotes for which it was intended. Instead, the compound turned out to be many times more lethal than thallium; a pound of it is sufficient to kill a million pounds of susceptible animal life. Worse, it is highly stable in rodent bodies and can thus cause secondary poisoning of the scavengers and predators of rodents. Not only the coyote but also the kit fox or the black-footed ferret that dines on a prairie dog killed by a particle of "1080" may also die. Thus rare and desirable animals, as well as the so-called pests, have become victims of an ill-considered policy.

Today the traditional system of predator and rodent control is being challenged. Scientists are beginning to wonder whether it is sensible to kill both rodents *and* their predators. Considerable government funds are being spent, for example, to exterminate coyotes throughout federal lands, even where grazing of livestock is not permitted. But Frank and John Craighead have studied the coyote in the sagebrush country of Jackson Hole, Wyoming, and found it to be a useful member of the wildlife community there.

According to trapping records, about half of Yellowstone Park's coyotes move down country (22 to 115 airline miles) in the fall and collect in the narrow valley of Jackson Hole, where large acreages of hay are grown for the herd on the National Elk Refuge. Big populations of mice in the hay fields attract the coyotes (20 were seen at the same time in one and one-eighth square miles). There they hunt until the blizzards of January kill some of the elk, whereupon there is plenty of carrion to eat. The coyote is a more effective rodent-control instrument than poisoned grain or traps. It is a significant part of the whole complex of rodent eaters in the valley—hawks, owls, badgers, bobcats, weasels, skunks, shrews—which together curtail the booming mouse population. In addition, the coyote joins a team of scavengers—ravens, golden eagles and bald eagles—in the important function of clearing away carcasses, as well as killing some of the sick and starving animals in this congested, big-game wintering area.

A Vanishing Predator
A black-footed ferret makes a rare appearance on a South Dakota plain. Disappearing with the killing off of the prairie dogs, this species is now almost extinct.

The secret of the coyote's success during the present era, in spite of "1080" and all the rest, is a combination of good survival instinct, big litters of pups, and the opportunity given to fulfill its traditional role on the prairies and plains. When rodents are plentiful, the coyote is a mouser. When they are scarce, because of close grazing and lack of food and cover, this persistent hunter turns to bigger game—and bigger game, for the coyote, can be sheep, calves, deer or wapiti.

6
The Desert and the Chaparral

Unevenly distributed rainfall, sharply rising temperatures during the day and loss of heat at night, and the "dust devils" and big wind storms stirred up by such extremes in climate—these are the factors that have created the sparse wilderness of the Southwestern desert valleys.

Low mountains of pink granite and black volcanic rock are eroded as fast as the infrequent but heavy rains can carry off broken rock, gravel, and sand. The silt from these deposits spreads downward in broad, gentle slopes called bajadas, to the flats. Unlike the forests and the grasslands, the desert is a place of constant erosion.

Here thick-leaved shrubs with deep roots and water-storing cactus plants with quick-growing surface roots have learned to live for months and years at a time with only a few hundredths of an inch of rain. Few plants are established, and these survive with a very low rate of reproduction. It is life held in suspension, as it were. Growing things pick up water and increase wherever possible, but sometimes go on without perceptible change for fifty or a hundred years. In a stand of saguaros, for example, observation over a long period shows that one plant matures for each one that dies.

On the mountain slopes, the desert gives way to thin bands of semi-arid vegetation—open grassland and woodland of evergreen oaks, chaparral and piñon pine. On the higher mountains of southern California, New Mexico and southeastern Arizona, the desert yields to pine and fir forests. The difference in annual precipitation between the mountain tops and the lowlands may be almost 30 inches.

The dense, interlocking thickets of evergreens called chaparral got their name from the Spanish word *chaparro*, for scrub oak, back in the mission-ranch era of California, and the leather overalls worn as protection for riding through such thickets are still called chaps. Nothing like the great chaparral region of southern California exists beyond the rain shadow of the Sierra Nevada, although a few of the live oaks and other shrubs found in California also grow on Arizona mountainsides.

Rainfall comes at different times of the year in the three major desert regions of North America. The Mojave receives the tail end of the Pacific westerlies in the winter. The upper margins of this desert are more or less delineated by the largest of the yuccas, the Joshua trees. The Chihuahuan Desert—a high tableland extending from Mexico into New Mexico and West Texas—receives rains from the Gulf of California in the summer. It is the country of succulent-leaved century plants and giant daggers. Most of the state of Arizona and the lower Colorado River Valley of California lie in the Sonoran Desert. Here rains may come in summer or winter. It is the lowest and hottest of the desert regions, dotted with mountains like wet islands in a dry sea. The Arizona uplands benefit from both winter and summer rains; in consequence, this is where one finds the greatest development of cactus—plants that are specialized for water storage, with leaves reduced to thorns and stems made into barrels.

In the Sonoran Desert, the giant saguaro stands sentinel over a varied and often luxuriant community of joint-stemmed chollas, smaller cacti, and shrubs. The thorny wands of ocotillo stay leafless for all but the brief rainy season. The green-barked paloverde has leaves, but they are tiny and barely visible through most of the year. In the spring, however, the paloverde becomes a glorious flowering of gold. The gray-barked iron wood and mesquite have slightly larger, symmetrically-arranged leaves with tough, shiny surfaces that minimize loss of water. The last three plants mentioned are tree-sized legumes and bear a rich harvest of seed pods on which wildlife feed. Their branches are often laden with mistletoe, a parasitic plant that produces berries on which many birds subsist before the winter rains bring out the ground cover of small annual flowers that attract insects in the spring.

At Gates Pass, west of Tucson, the saguaro forest sweeps up to the topmost ridges, its widely spaced, fluted columns reaching heights of 30 and 40 feet. Here one can see the plant's preference for a south-facing slope. It has always been thought that the gentle, prolonged winter rain was more vital to the giant cactus than the quick run-off of summer thundershowers. But in the hottest season the saguaro adds most to its vegetable bulk, quickly putting out root hairs to catch

the rain whenever it comes, and piping it up the bundle of hollow ribs that hold the stem erect. The pleated surface swells and contracts to the rhythm of water intake and output. If by any accident the column is broken, lateral branches start from the wound and curve upward toward the sun. Strange, twisted formations result from the dry years that constrict these arms, turning them down, and the return of moist weather that stiffens and turns them up again.

Fortified with stored water, the saguaro blooms unfailingly each year, the flowers on its crown and on the tops of its arms usually opening in May, in advance of the summer rains. In the latter part of June or July, the filmy, white-rayed corollas are replaced by fig-shaped fruits that burst open when ripe, revealing a full-seeded, crimson pulp.

As the base of the bajada is approached, the cactus forests become confined to small drainageways, and the intervening spaces are filled by bursage and creosote bush, which dominate the desert basin. Creosote, the only plant common to all three deserts, sends down a tap root and has surface roots as well. Its waxy, dark green leaves both conserve water and absorb moisture from the air during the cool desert nights. Because the fine-textured soil does not absorb much water, the plants space themselves to cover only 15 to 20 per cent of the surface. Under frequent sheet flooding, they grow as high as six feet, but they are usually only about two feet tall. Recently biologists have discovered that creosote bushes deposit in the ground a toxic substance that assures them plenty of room by discouraging the growth of other shrubs. After soaking winter rains, however, the topsoil may be covered for weeks by the dense growth of a tiny plantain with a large seed head— the desert Indian wheat—and by grama grasses.

Perennial heat and drought have prevented desert plant growth from flourishing in the basins, except along deep arroyos. Here in the gullies,

A Saguaro Giant
At the base of a sloping bajada, an old saguaro stands as stark and lonely as a nearby peak in the Ajo Mountains of Arizona. The big cacti are scarce in hot lowlands.

217

mesquite and blue paloverde send their deep roots down to find the permanent water table. As the heat of noon descends on the Sonoran desert in May, the arroyo is silent except for the buzzing of bees in the golden paloverde blossoms, and inactive except for the processions of harvester ants carrying away fallen petals. Shade is at such a premium that the occasional human intruder—tourist or cowboy—who finds a bit of it will most likely have to chase away a jackrabbit to make room for his own siesta.

The rocks in the wash give shelter to a predator of the ants, the regal horned lizard. On the branches of the paloverde overhead, a dove family may roost. Every animal that lives here obtains its water in a different way. The ant can live off metabolic water released by the burning of carbohydrates in its own body. The lizard utilizes the body fluids of the ants it consumes. The antelope jackrabbit eats cactus plants that contain stored moisture. At the end of the day, the mourning dove flies to a mountain water hole to drink (but it also derives a certain amount of fluid from the nectar in cactus blossoms).

The jackrabbit's out-sized ears not only improve its hearing but dissipate unwanted body heat. Exaggerated spines on the cold-blooded lizard serve another purpose—defense. Lacking the jackrabbit's speed, the little lizard has evolved in imitation of the cactus plant. A crown of well-developed thorns around its head can choke the rattlesnake that tries to swallow it.

The Biggest Ears
In the shade of a prickly pear, an antelope jackrabbit waits for sundown, which is its time to feed. Bigger than other "jacks," it also has much larger ears. This is an advantage in the desert; by increasing body surface and thereby increasing potential heat loss, its oversized ears help to keep the animal cool.

The Most Horns

Living in the hottest desert area of the U.S., the regal horned lizard of the Sonoran sands has evolved an unusual cooling system. Its body surface is increased by twice as many occipital horns as possessed by its relatives, the so-called horned toads; that is, it has four horns instead of two. The knobby appendages, positioned like a crown around the head of this dragonlike reptile, give the big lizard its "regal" appearance.

A Tiny Snake
Only nine to ten inches long, the banded sand snake of the Sonoran Desert has a pointed snout shaped for swimming in sand. Most of the time during daylight hours it cruises just under the surface in sandy flats and dunelands.

A large, heavy-bodied, venomous lizard, the gila monster, also lives in and around the desert wash. Its red and black beaded skin advertises a neurotoxic poison similar to that of the coral snakes. When this reptile catches a prey animal, it chews with grooved teeth through which its venom flows until the victim's breathing and circulation stop. According to records, a total of 136 people have been bitten by gila monsters, and 29 of them died. As our only poisonous American lizard, the gila monster interests more persons than it endangers. Unlike the aggressive sidewinder, it never attacks unless it is molested.

Away from the shade, the food, and the intermittent flow of water provided by the arroyo, life goes underground. Internal adjustments for the economical use of water become more important.

A Mexican reptile, rare in our part of the Sonoran desert, is the banded sand snake. Its adaptations for a sand-burrowing existence include a streamlined head, a strong snout, a non-retractable neck, and a countersunk lower jaw that keeps sand particles from getting in its mouth. Smooth body scalation and angular "treads" on the belly insure traction with the least amount of friction while it is crawling through the sand. Only at night, when it is actively searching for ants, grasshoppers, spiders and beetles, does

this snake sometimes come to the surface. In open areas between scattered desert bushes, as well as beneath them, its movements can be traced by a maze of tracks where the sand has caved in behind its burrowings.

Most of the desert mammals are small, nocturnal, and physiologically adapted, like desert shrubs and trees, to conserve the water in their bodies. In lieu of a reservoir such as an oversized root or stem, the pocket mouse, kangaroo rat, and antelope ground squirrel all have modified kidneys that reabsorb two to three times as much water from the urine as those of a laboratory white rat. The antelope ground squirrel tolerates outside temperatures up to 107 degrees F. When the mercury climbs higher—often to 115 degrees F.—the animal retires to a burrow and washes its face and shoulders with drool.

Larger, less active, and more placid in temperament, the round-tailed ground squirrel has none of these adaptations except a fairly high tolerance of heat. During the dry season, this animal simply goes into estivation for months, like its relatives in the mountains. The little pocket mouse, living under the same conditions, goes into estivation for shorter periods.

None of the desert creatures combines so many adaptations as the kangaroo rat. The big-eyed little rodent with the long, brush-tipped tail has

A Gentle Monster
*Feeling along with a sensitive tongue, a gila monster
raids a quail nest. The venomous lizard, which grows up to
sixteen inches in size, is protected by Arizona law.*

A High Jumper
*Despite its name, Merriam's kangaroo rat is really
a big jumping mouse. Nocturnal and solitary in habit, it
feeds on a variety of seeds and sparse desert vegetation.*

the ability of the ant to increase its water content on dry seeds alone. (One gram of carbohydrates in the seeds produces .6 gram of water when oxidized; fats are similarly converted and proteins.) The rat's kidneys also conserve water. The two adaptations combine to maintain the animal's water balance on a dry diet—as long as it comes out only at night, when the relative humidity is greater than ten per cent and the temperature of the air no more than 75 degrees F.

A nocturnal animal that preys on the rodents is the small hooded skunk, a primarily Mexican species with a ruff on its neck and a longer tail than the common striped skunk (which originally came to North America by way of South and Central America). The hooded species has omnivorous habits, similar to its larger relative, but ranges no farther north and east than the limits of the giant saguaro.

In years when the woody shrubs, because of lack of rain, do not put out leaves or flowers, the water-storing saguaro and smaller varieties of cactus are vital to the animal community. Insects taking nectar from the cactus flowers are eaten by flycatchers and by the gila woodpecker. Ripening cactus fruits are eaten by other birds, including doves. Fruits that fall to the ground are eaten by rodents and coyotes. The seeds are carried away by ants. The production of seeds is more than adequate for wildlife, and also for the Papago Indian harvest. It has been estimated that a mature saguaro produces twelve million seeds for every seedling that takes root to perpetuate the stand.

The seedlings that survive are usually in sheltered places, among bushes or rocks where they are protected from frost, sun and washouts. They mature very slowly, and at ten years of age may be only two inches tall; but as water storage increases, the young saguaros quickly attain the size of a large barrel cactus. It is not unusual to see several of them growing right up through the branches of a paloverde or other "nurse plant." Once well established, they may continue to grow for 150 to 200 years, resistant to drought and to the cuts made by desert animals.

Small rodents gnaw spirals in the saguaro's base, and sometimes creep up and riddle its arms with their teeth marks. Woodpeckers drill countless holes, in which they and other birds raise their young. The plant repairs these incursions into its pulpy flesh by lining the wounds with woody fibers. But when it becomes old, the towering saguaro shrivels under each new attack. And it is now particularly susceptible to the burrowings of moth larvae that carry bacterial necrosis. Gradually the dried flesh falls away from the wooden skeleton, until finally this bundle of sticks crashes to the ground like an old tree.

When conditions do not favor new growth in a saguaro stand for twenty or thirty years, a great age span develops between the large, old plants dominating the slopes and the young ones just overtopping the brush. Growth may be retarded by the shifting local weather patterns that distribute rainfall sporadically over the desert, or by large numbers of cattle that reduce the ground cover of grasses and shrubs and trample the seedlings of the big cactus. As the grass and the saguaro "nurse plants" disappear, prickly pear and creosote bush invade, and populations of rodents increase—especially if coyotes and other carnivores have been killed off by predator-control operations.

The old Saguaro National Monument, east of Tucson, lies on sandy bajadas of the Rincon Mountains. Here drought and overgrazing early in the century contributed to a die-off of shrubs, to slope erosion, and to an increase in wood rats and ground squirrels—all leading to the decline of the aging saguaro stands. Warned that there might be none of the plants left by the year 2000, the National Park Service built a fence to exclude cattle from the "loop drive" of saguaros that most visitors see. Since the fence was put up in 1958, rainfall and protection from grazing have together replenished the ground cover (although some shrubs have not returned), and young, healthy saguaros of various sizes dot the landscape.

The old monument, however, may never produce as luxuriant a forest as the park's new section on the rocky slopes of the Sierra Tucson. The stands there are younger, and fifteen- or twenty-year-old saguaros cover the area. Comparative photographs, taken in 1908 and 1962, show that the density of the giant cacti and paloverde has not changed during this period and that the pygmy forests of cholla have spread. There has been no cattle grazing and no coyote control

Back with the Kill
*In the cool of the evening, when it prefers to hunt,
a hooded skunk carries a kangaroo rat to its den. Like the
striped skunk which it resembles, this desert animal
has a choking, blinding spray to use for self-defense. But
its self-assurance because of this can be fatal. The
skunk has poor eyesight, and often fails to take notice
of its own predators—owls, coyotes, bobcats and automobiles.*

A Cholla Nest
*Deep among the spiny branches
of the chain-fruit cholla,
a roadrunner incubates her eggs.
This cactus, which grows up to 12
feet high, provides fortified nesting
places for many desert birds.*

A Cactus Forest
*Silhouetted dramatically against
an evening sky, saguaros
stand like tall sentinals in the
Sonoran Desert, towering
above the smaller cacti and shrubs
that grow in the cactus forest.*

for thirty years, and—most important—soil conditions and local weather have been favorable.

The teddy-bear cholla has formed extensive hillside gardens that are deceptively soft and silvery in appearance, considering their dense armor of spines. On one plot of ground, 21 of the joints that fell around the bases of the cholla plants had taken root—increasing the stand from 30, counted in a 1908 photograph, to 51 today.

Sometimes covering many acres of bajada, the Chollital (place of the cholla) is a favorite nesting area of birds. The chollas are related to prickly pears and other low-growing cacti. They are shrub-sized and flower modestly. What distinguishes them particularly is their fruits; in the chain-fruit cholla, these grow one out of the other towards the ground. Within this thorny world of fruits hanging like Christmas tree ornaments, cactus wrens build nests in the shape of a football.

Ornithologist Herbert Brandt has described the construction of one record nest, two feet high and 18 inches wide, which he named Casa Grande. Into it the wren architects had woven a gallon of various bird feathers they had collected

from near and far—the glaucous feathers of the white-winged dove and the ornate feathers of Gambel's quail, the gilded flicker and one of the orioles. There were some 10,000 fine grass stems, perhaps 500 coarser stems of taller plants, and 500 strips of bark. If placed end to end, all of the bits and pieces of vegetation, fur and feathers would have stretched a full mile.

Most wrens are not so ambitious, but their nests always offer considerably more protection from the elements than a simple platform. Other birds, including house finches and doves, often renovate and use parts of the wren nests after they have been abandoned by the builders. Because the breeding season is very long, the birds of the Chollital normally raise more than one brood; young can be found in their nests all through the summer.

The handsome, crimson-breasted house finch is perhaps the scrappiest of the cholla meadow's occupants; it even chases the wren out of its home territory. This bird builds a nest of fine weed stems, tucked in among the cactus spines. When the young finches are ready to fly, some of them may be impaled while trying to leave the nest.

224

Home in the Cactus

A young cactus wren peers from the door of its home, a huge ball of finely-woven, grassy materials. To enter, the parents must squeeze through a narrow passageway before reaching the inner nursery.

A New Tenant

Safe among thorns, a mourning dove covers her eggs, laid in the ruins of a wren's abandoned nest. Doves may build their own nests. In any case, they are careful to pluck away the most dangerous of the surrounding thorns.

A Desert Tragedy
Proof that the protective spines of the cactus are sometimes deadly,
a half-grown finch lies skewered on a cholla branch. The
bird had left the nest before its wing feathers were fully out.

However, most of the bird parents in the Chollital nip off the dangerous tips of the cactus spines, making it safe for both adult and young birds to fly in and out.

Protected as the birds seem to be in their thorny fortresses, the eggs and young are frequently stolen by climbing snakes, which persevere despite the danger of being skewered. No kind of nest construction excludes them; a racer five feet long can squeeze through the tiny entrance to the wren's home. And the bull snake works its way patiently up past all obstacles in the paloverde and the cactus, to raid the eggs and young of any species it can find.

Above the cholla, high-rise saguaros are full of gilded flicker and gila woodpecker apartments. As soon as these are abandoned, they are "leased" by other birds returning from their winter dispersal into the Sonoran lowlands—elf owls, screech owls, purple martins, sparrow hawks and flycatchers.

For colonies of birds like purple martins, the thoroughly punctured saguaros around Tucson are just as ideal for nesting as the tall pine stubs in the Chiricahua Mountains of southeastern Arizona, where nesting holes have been carved out by acorn woodpeckers. The giant cactus therefore extends the range of the martins, in a discontinuous fashion, from the high mountain zones directly into the desert.

Finally, in the crotch of an old saguaro the red-tailed hawk makes its substantial stick nest. In the heat of midday, the downy white young are shaded beneath the mother hawk's spread wings, and for six weeks—many weeks longer than the smaller birds of the saguaro community—they will be fed and cared for in the nest. Both parents hunt the desert until dark every evening and again in the cooler hours of the morning, searching for rodents or snakes. As long as the hawks are tenants, no other bird will live in that particular cactus. But woodpeckers in neighboring saguaros often will join together to "mob" the predator.

The hawk's nocturnal counterpart as a bird of prey, the great horned owl, often claims the hawk's abandoned nest. Or it may find a niche in some small mountain cave, close enough to visit a spring or a pool at night. Under its nest is a tell-tale pile of bones, representing a cross-section of the most abundant animal life in the locality. Hawks and owls usually do not digest bones, but eject them in little pellets of fur called castings, thus leaving a record of what has been eaten: rabbits, pocket gophers, mice, kangaroo rats, three kinds of skunks, bats, and even horned lizards. An owl's castings might also contain a large number of the legs and shells of beetles, grasshoppers and other insects, and occasionally the remains of a small bird. No other bird of prey in the desert is so bold a hunter—and none is less particular about what it eats. If the hawk should cross its path at twilight, the great horned owl may attack and eat the predator. (On the other hand, if the owl comes out during the daytime, it may just as likely be eaten by the hawk.)

A strange hibernator, the poorwill "sleeps" the winter out in the hollow of a saguaro, or under the pineapple-like leaves of the century plant, or among canyon rocks. Hopi Indians, early Mormon settlers and Arizona prospectors all knew about the unique behavior of this bird, but their tales were discounted until, on December 29, 1946, biologist Edmund Jaeger found one hibernating in a cold canyon bottom. In late November of the next year he found the poorwill again "dead to the world" in the same place. Testing its internal temperature, he discovered this to be 64.4 degrees F. instead of the 106 degrees F. normal for a bird. When he returned in February, the poorwill had lost very little weight. It woke up while being banded and flew away. But since it now carried an identifying band, Jaeger could be certain that this same bird occupied the rock niche for a third season in 1948 and a fourth in 1949. Since then, several other hibernating poorwills have been discovered, including one which was brought in to the Desert Trailside Museum, north of Tucson, in 1953. During its five-day stay there in an unheated adobe building, the bird's temperature dropped every night to 55.7 degrees F., and then rose somewhat every day.

One hundred and twenty-five million years ago, sediments of silt, sand and sea shells were accumulating on an ocean floor in the present Southwest of the United States. Then, under stresses and pressures from within the earth, the land rose and cracked into great blocks. Some were thrust up to form mountains and ridges, and others became valleys. At the core of the disturbances was molten lava, some of it plugged below the surface and some escaping to flow out over the landscape.

Millions of years of erosion of this region have stripped away soft layers of sandstone and limestone, leaving today's panorama of low, castle-like formations—hard granite and black lava plugs that are isolated outcrops on the undulating, arid plain. Pinnacles stand stark above the slanting apron of crumbled rock that surrounds them. On such slopes little vegetation grows; without trees to give them scale, the jagged ranges seem higher than their actual 4,000 to 5,000 feet, and as flat as cardboard cutouts. It is the gathering of summer clouds about them that gives the peaks the visual dimension of real mountains. At sundown, the pink granite is turned into flaming orange against a leaden background. At this season there is always the threat of a downpour miles away that will bring raging torrents down the arroyos which criss-cross the desert valleys.

This was the look of the Kofa Mountains, north of Yuma, Arizona, as we drove over rough trails to the inner canyons for the annual bighorn sheep count in June of 1966. The Bureau of Sport Fisheries and Wildlife counts the sheep at the driest and hottest time of the year, when the animals must come in to a few mountain seeps and water holes to drink at least once in five days.

The trail led from numerous lowland crossings of arroyos to one main creek bed that was the road to Tunnel Springs. Only occasional saguaros stood along the way, their red fruits ripe and open. With altitude, there appeared sparse colonies of century plants (agaves), some of them as tall as twelve feet. They were in full bloom, and were everywhere attended by darting, hovering,

The gila woodpecker arrives at its nest hole with an insect for its young.

Living Accommodations
In the saguaro forest, a gila woodpecker (above)
has its nest in the hollow of a big trunk.
The nest hole is drilled in the fall, so that the
cactus wound will heal and dry up by spring.
A red-tailed hawk nests out in the open (below).

Week-old red-tailed hawks wait for their mother in a saguaro crotch.

Snakes are tidbits for baby hawks.

desert hummingbirds. We camped below the springs, located deep in a mountain cave. The drinking arrangements had been improved by the installation of an old-fashioned bath tub, a man-made water hole.

Near the mouth of the cave, a blind had been set up among a dense growth of dwarfed oaks and other shrubs. Waiting in the blind for the appearance of sheep, we spent the hours counting the population of nesting birds—mourning doves, Gambel's quail, blue-gray gnat-catchers, ash-throated flycatchers, hummingbirds. The identity of one young bird in dull plumage puzzled us until the brightly-colored male Scott's oriole came in to land on a branch just in front of the blind to feed its offspring.

The traffic into the oasis cave was continuous. We were surprised to see quail *climb* to the water hole. Instead of flying, they hopped up the rock surface, with big flocks of young following a few adults. We counted 69 young in two broods. For quail in wet years, when the grasses and annuals and the insects on which the chicks begin life are all abundant, there is a second nesting. (In dry years, on the other hand, quail may not nest at all.) In mid-June, the first young of a wet season are given over to the care of "babysitters" while the majority of the adults nest again. There by the cave we seemed to be witnessing a sort of avian kindergarten. A male in charge of one brood was seen to descend the rock and try to chase away another brood. Since females are relatively scarce, in the mating season there are many surplus males, the losers of arena-like battles for supremacy. They call hopefully but in vain all

through the spring for a mate. Could the baby-sitter role be one for the unmated cock?

As though they had been waiting in the wings of a stage, the bighorn sheep came suddenly into view, filing in small bunches to the spring in the cave—the ewes with their lambs, the rams in bachelor parties. The sheep counter can recognize each animal by the horns, physical condition, size, and marks on its body. On the third day of our watch, the dryness and heat brought in 13 sheep by mid-morning. (It was 90 degrees F. with only six per cent humidity at 7 a.m. that day.) The proportion of young to adult animals is a good indicator of the success or failure of a given year's mating. At Tunnel Springs, where we watched, there were many lambs four to six months old, all males.

Ewe and yearling, seen at close range through the binoculars, had short, semi-curved horns and elongated, Picasso-like faces. Rams three to four years old wore the half curl, and older ones the full curl. The refuge warden told us that the "annual rings" in the horns are not always dependable clues to age. In some years there are two dry spells when food is scarce and growth of the horn is arrested; consequently two ridges are formed instead of one. Since it was too early in the season for the butting contests between rams, we saw only preliminary play between some half-curls—a squaring-off at the place of descent and a resounding bang of foreheads. Then the two would jump down, one behind the other, with a half-falling, half-leaping motion, to land gracefully at a lower level on cushioned hooves.

The bighorns are highly nervous at the water hole, where they know danger may be waiting. They often stop and cough before entering the cave, as if testing to see whether anything will move or make a sound in the shadows. Lambs younger than four months old may be left on the mountain top for safety while their mothers descend to the spring. We noticed that some animals entered the cave but never lowered their heads to drink. Interrupted by another sheep, or perhaps frightened by the scent of a coyote or bobcat, they would leave still thirsty.

Lack of water has done more in recent years to keep down the number of sheep than have predators or poachers. With the improvement of the natural water holes, or "tanks," and of small springs and seeps, there will some day be a dependable water supply in every square mile of the Kofa Mountains. At present, the 1,000-square-mile game range may support as many as 300 bighorns. The Desert Game Range in Nevada, which is larger than the state of Rhode Island and encompasses six mountain ranges, has 1,000 sheep (and another 1,500 have wandered outside of the protected area). In the Mojave, scattered bands total not less than 1,500, while Cabeza Prieta has at least as many as the Kofa.

Since protection of the animals began in the late 1930's, the bighorn sheep have increased in the arid ranges; there are, in fact, more of this desert race than of the mountain bighorns that dwell in the high Rockies and Sierras. Since the 1950's, hunters in Nevada and Arizona have been allowed to stalk some of the herds when they decend to the valleys for the winter.

After the rains, in early spring, the desert bighorns return to upland canyons that are covered with wildflowers. One of their favorite foods is lupines. Later in the season, they may eat Simmondsia leaves and nuts, the leaves of cat-claw, young twigs of paloverde, Indian wheat, and grasses. John Muir has described the remote ledges on which the bighorn drops its young in the high Sierras. There are few records of birth in the desert, but apparently the young are dropped in a place no more remote than the canyon floor, mostly from December to May—although the ages of lambs counted at the water holes indicate that some births occur in nine out of 12 months.

Like the quail, the bighorns are dependent on the rainfall that once in a decade or two fills the "tanks" and playas, brings out desert gardens of colorful annuals, and provides a rich harvest of seeds. Because they can store water or can remain quiescent, the perennials are more stable, less likely to reflect change. More than any other growing thing, perhaps, the century plant is symbolic of the desert drought cycle. This plant spends its entire life manufacturing and storing the energy to send up one great flowering stalk that may grow as much as 12 inches in a single day. Once the seeds have matured and blown away on the desert winds, the plant dies.

Though the desert is a checkerboard of local climatic effects, with plants declining in one place and flourishing in another, there are indications that the basins as a whole are becoming hotter and drier—thus causing desert flora to migrate up the mountain slopes, where moister and cooler conditions prevail. All the deserts of the world have warmed slightly since the 1880's. At the Phoenix, Tucson, and Yuma weather stations, the increase in mean annual temperature amounted to two degrees F. between 1898 and 1959. Together with the over-all temperature increase, there has been an average decrease in precipitation in Arizona and western New Mexico of about 25 per cent since 1921. The combination of greater heat and greater aridity seems to be making a more desolate area of the lowlands—not only where cattle have grazed in large herds since the Spanish-mission era, but also in the interiors of the Pinacate craters, along the border between Arizona and Mexico, where nothing has disturbed the vegetation.

There is evidence that the present pattern of arroyos and severe flooding with its attendant erosion dates from about 1870. Before the Civil War, the San Pedro, the Santa Cruz, and their tributaries—the Gila and the Sonoyta Rivers—wound sluggishly along for much of their courses through grass-choked valleys dotted with cienagas (swamps) and pools. In the streams, salmon trout and beaver were abundant, and turkeys hatched out in the waist-high grass. Today this kind of environment can be seen only around an oasis such as Quitobaquito, a one-and-a-half-acre pond in Organ Pipe National Monument, frequented by water birds, dragonflies, and doves.

Dangerous in a Corner

A bobcat snarls and lifts a menacing paw when cornered by an enemy. This predator feeds largely on small game, but in the hot summer months stays close to water holes in hopes of catching a bighorn lamb.

The 1880's brought earthquakes and extreme drought, followed by catastrophic floods that cut deep channels in the plains. The water table fell, and rivers disappeared under the desert sands. Except in their upper reaches or in places where rock formations force water to the surface, the river beds are for much of the time dry, sandy wastes that support little, if any, vegetation. Five to thirty feet above the waterless channel, there is usually a line of mesquite; the 40- to 60-foot-deep roots of this shrub need the aeration of a dry bank and are able to reach permanent water down below.

While floods cut deep arroyos, 315,000 cattle (according to an 1895 census) were denuding the grassland south of the Gila River. The cropping of grass helped to upset the balance between water infiltration and runoff. With more runoff, the tough-coated seeds of the mesquite were more effectively carried from place to place, later to break open for germination. Fewer fires swept the landscape, and this, too, favored the shrubs. Is the pattern of shrubby valleys and vast creosote stands irreversible? Or are we in the midst of a long-term cycle in which trenches are cut in the floodplains and then gradually leveled by erosion until marshes again appear? The climatologists are not certain.

Meanwhile, the saguaro and its shrub community are dying in the lowlands, desert plants are moving up the mountainsides to invade the grass areas, and at higher elevations there is a general die-off of oaks. Both grass and oak communities appear less open than 40 or 60 years ago because of the mesquite invasion.

Organ Pipe National Monument and Cabeza Prieta Game Range together tell a detailed story of the cactus desert—its past, its present, and its probable future. The topography includes higher mountains, such as the Ajos, and a lower series of parallel ranges that seldom top 3,000 feet but are extremely rugged, such as the Sierra Pinta, where bighorn sheep trails lead almost vertically down into Hart Tank. There is upland saguaro and paloverde intermixed with the Mexican organ pipe and senita, and lowland saguaro rising in lonely splendor here where the lava fields of the Pinacate spill over the border, sand dunes blow in from the Gulf of California, and creosote

stretches to the far mountains. Rainfall drops from about eight inches a year on the eastern border of the Game Range to about four inches in the Yuma desert to the west.

It is this panorama that would make up the proposed 1,242,000-acre Sonoran Desert National Park, the seventh largest unit in the U.S. park system. The dry western reaches, set aside for the bighorn, are nearly trackless except for the ruts left in lava by Spanish wagon wheels. To strike out across this wilderness from the east is to know the feeling of blazing a trail. The only constant landmark is the pink granite mountain with the black lava cap, Cabeza Prieta, near the town of Yuma to the west. From Father Kino, who first made the journey in 1698, and the gold-rush settlers who named it "the Devil's Highway," down to the present day, the story has been the same. Each individual traveler makes his own way through the desert, keeping an eye on the distant mountain. And as he goes, he is reminded of defeats and victories of man in an arid land. An old coach road along the Mexican border still bears the sign "60 miles, no water." (No one has ever bothered to complete the fence along this border, which is crossed mostly by stray cattle and game.) The grave of a prospector is marked with an iron cross. The well where a ranch used to be is a welcome oasis. The sound of distant firing on the site of U.S. gunnery and bombing ranges reminds the traveler of the most current use to which the desert is being put.

It is not surprising that the conditions which cause us to shun the desert do not trouble the bighorn, which seeks solitude, avoiding other animals and people in the most desolate and hottest places. The bighorn has ready-made shelters in the wind caves of the mountains, to which it has always retired to escape the heat of the day.

Gone are the Sand Papagos who painted the walls of those caves, ground the Indian wheat with stones, and hunted with arrows, driving the sheep into ambush at the tops of pinnacles. Gone, too, are the cattle that fattened on the Indian wheat and grama grasses—and the cattlemen. If a concrete road is built as far as the water hole at Hart Tank, as proposed, it will make little difference to the sheep. They are regular visitors at the water hole only when the temperature reaches 110 degrees F. in July and August, and in this heat

The Organ Pipe

Reaching to sixteen feet, the organ pipe cactus sends up multiple columns from its base, and also grows arms like those of the saguaro. Since the plant is extremely sensitive to frost, it is found no farther north than the Sonoyta Valley in Organ Pipe National Monument.

The Senita

Just about ten feet in height, the senita ("old one" in Spanish) is crowned with clusters of long gray spines that resemble whiskers. The cactus—abundant in Mexico—can be seen in the U.S. only along the extreme southern edge of Organ Pipe National Monument.

35

the number of people who will climb the sheer cliff at the end of the canyon to spy on bighorn sheep will be small indeed. The modern tourist, like other interlopers before him, is no more than a transient disturbance.

Scattered mountain ranges that stretch from the southern tip of Nevada across southern California halt the advance of the saguaro into the Mojave Desert. From 1,000 feet at the eastern end of Joshua Tree National Monument, the altitude increases to nearly 6,000 feet in the Little San Bernardino Mountains. These highlands are marked by outcrops of granite-like quartz monzonite, which has cracked into huge boulders, the so-called "jumbo rocks." The plants that replace the saguaro here are all yuccas. On the gently sloping mesas and in the rough terrain at the base of the mountains grows the Mojave dagger, and

from this level up to three or four thousand feet, the Whipple yucca. In some places these plants give way next to the Chollital, but the slopes above 3,300 feet are dominated for the most part by the distorted forms of Joshua trees.

The head of a community of drought-resisting plants, the Joshua tree (named by the Mormons for its resemblance to someone praying) does not bloom every year. The seedling grows straight up, producing a single spray of needle-like leaves, then another, and another, building in every direction possible. The result is a trunk with dozens of parts, having as much direction as a child's pop-on toy, double and triple-jointed. In the wetter years, blossoms appear in a cone-shaped bunch on the top of each spray. As the needles die, they turn downward, harden and become sanded by the windstorms of the desert into a rough bark. Young Joshua trees are shaggy, but older ones are

Joshua Trees
A young Joshua tree growing in rocky surroundings in Joshua Tree National Monument (below) *resembles a variety of the yuccas found in California desert valleys. But after many years the plant will turn into a tree with a Medusa-like crown of twisting branches* (right).

A Big Reptile
The biggest of the lizards in the Mojave and Sonoran Deserts, the chuckwalla may grow a foot and a half in length. At night the animal becomes inactive, wedging itself into a rock niche. After warming up in the morning sun, it forages on flowers, fruits and leaves.

protected by this bark right up to their spreading crowns. The ladder-back woodpecker drills into Joshua trees, leaving old holes to be occupied by screech owls. The decaying stubs house numerous rodents and insects.

Like all yuccas, the Joshua tree is dependent for pollination on one particular yucca moth which has specially developed mouth parts for collecting balls of pollen. The moth flies around carrying the pollen until she finds a yucca blossom suitable for egg-laying. Then she punctures the ovary of the flower with her ovipositor and inserts her eggs, one by one. There are usually about six. After each egg is deposited, the moth climbs to the top of the pistil and rubs part of the sticky pollen from her mouth onto the open end of the stigmatic tubes. This fertilizes the plant and assures a supply of developing ovules for the larvae of the moth, which will eat some of the seed capsules and leave others to mature normally. When fully grown, the larvae break out of the seed pod and drop to the ground, where they may remain more than a year. After the soil has been softened by a spring rain, the moth pupae pull themselves to the surface with undulating movements of their bodies, and the cycle is repeated. The pupae do not emerge unless there is rain enough to produce blossoms on the yucca.

The lank, deciduous smoke tree is another member of the shrub community along the washes. In June, its bare branches and the brown tendrils of the coyote melon, creeping over the ground, give the Mojave (which gets no summer rain) the look of a garden in autumn. The dried bean clusters of paloverde, mesquite and ironwood rattle loudly in the late afternoon wind that rises to erase the sign of animal tracks in and out of the arroyo.

Round, deep-sunk, plodding footprints, as of a Lilliputian elephant, belong to the desert tortoise that builds a cool burrow in the bank. The female of the species usually lays eggs at this season. The

238

shell of the egg, like the shell of the tortoise, is resistant to water loss.

Lizard trails in the desert are just dotted lines, etched delicately as the lizard's tail is dragged over the crests and through the shallow valleys of the sand ripples. If such a trail ends very abruptly under a creosote bush, the chances are its owner is a pale sand lizard that has dug itself straight down with its broadly fringed toes. Other lizards use the complex of holes dug by kangaroo rats; the little reptiles can be seen racing from bush to bush, sometimes accompanied by a convoy of noisy shrikes. Perhaps the largest of the speedy lizards is the desert iguana, which skims out of sight before it can be more than glimpsed.

Another lizard of the Mojave, the chuckwalla, has devised a unique means of escape from enemies. It shelters in the crevices of rocks, inflating the loose folds of skin on its neck and on the sides of its body so that it becomes wedged tightly in a narrow space and cannot be pulled out.

Lizards that live in holes in the ground often plug them up at night, for defense against the prowlings of various snakes. The sidewinder, the speckled rattlesnake, and the Mojave rattlesnake are the three desert representatives of their poisonous tribe. All eat kangaroo rats, pocket mice, and lizards, usually in that order of preference.

Snakes and owls are among the few predators that will eat the tiny, musky gray shrew of the desert bajadas. This animal has developed a larger ear than any other shrew in North America, probably in response to the same laws of heat regulation that have produced bigger ears on desert rabbits and larger horns on desert bighorns. It needs no water except what it gets from hard-shelled insects; and despite its habit of eating three-quarters of its own weight in insects daily,

A Tiny Mammal
The smallest mammal of the Southwestern deserts is the gray shrew. It breathes faster, has a faster heartbeat, less control of body temperature, and needs more food in proportion to its weight than larger animals. So the shrew alternately feeds and rests under piles of old cholla branches, as shown here.

it can go without food for one or two days with no apparent ill effects. Weighing only 3 to 3½ grams, it can occupy a beehive, using the old entrances made by the bees, but its usual home is under a pile of rubbish—cholla skeletons and dried agave leaves.

A bird perpetually in a hurry, the road-runner is as much a part of the Mojave as of the Sonoran desert. Early in the morning or late in the afternoon, this ground cuckoo patrols the countryside with raised crest and piercing eye, looking for some sign of a bird's nest, of mouse or lizard activity, or of snakes on the prowl.

The roadrunner has become an expert in dispatching snakes, and it shows considerable judgment in deciding which ones it will attack. At the Arizona-Sonora Desert Museum, we recently filmed the behavior of several roadrunners which were offered poisonous and non-poisonous snakes. They would not touch a rattlesnake two feet or more in length, but would kill and eat some very large non-poisonous snakes (the lyre snake, for example). Their *attitude* towards a poisonous snake was of a special nature. The attack on a sidewinder began with a display of wide-open wings and fast footwork, the idea being to make the snake strike, miss, and tire itself out. The bird continually stabbed at the snake's head with its long, sharp bill. Unable to crawl or sidewind away from the fast-moving attacker, the victim would stop striking and try to shield its horned head under its coiled body. But the bird always won, finally grabbing the snake below the head and whipping it relentlessly against the ground. Assured that there was no more life in the limp body, the roadrunner would drag it off into the bushes and swallow it whole at leisure. Horned lizards are also devoured after a great deal of battering —which seems to be the equivalent of tenderizing a tough piece of steak.

From Salton View, at an elevation of 5,185 feet, on the western edge of Joshua Tree National Monument, you can see the meeting place of

Snake Killer
After a battle to the death, a roadrunner holds the limp body of a rattlesnake in its bill. This lanky bird of the Southwestern deserts is expert at dispatching snakes.

240

Tenderizing a Meal

*For a roadrunner, a dead 18-inch sidewinder rattlesnake is just a piece of steak too large and tough to swallow. The bird handles the situation by grasping the prey firmly in its beak and beating it on the ground repeatedly until satisfied of its tenderness. The process may take as long as ten minutes. Then the roadrunner spends a whole half hour swallowing the snake, in slow gulps, until only its rattle hangs from the bird's mouth (**opposite page**). The digestive juices at work in the stomach of the roadrunner are so powerful that even the snake's bones and scales will be dissolved.*

Chaparral
Impenetrable thickets of red-barked manzanita spread
endlessly across the lower slopes of Big Pine Mountain
in southern California's Los Padres National Forest.

The Ringtail
A small raccoon with a foxlike face, the ubiquitous ringtail ranges through all three U.S. deserts and also the California chaparral. It does its hunting after dark.

the two deserts and the chaparral of the California coast. Below lies Thousand Palms oasis, thriving on water forced up by the San Andreas fault. West and northwest are coastal mountain ranges whose lower limits are covered with California sagebrush and chaparral. (The Santa Rosa Mountains are part of a system ending in Baja California, Mexico; the San Bernardino Mountains join a complex of ranges that end in the offshore islands of the sea lion colonies.) On a clear day, the Salton Sea—a desert sink filled by overflow from the Colorado River—is visible 30 miles to the south. This is the Lower Colorado, its plant and animal life more closely related to the Sonoran than to the Mojave Desert.

From here you can also see San Gorgonio Pass, in the San Bernardinos, where winds of sandstorm intensity link the cool Pacific lowlands around Los Angeles to the heated desert. On the other side of the mountains is the mild winter rainy season and abundant year-round sunshine that created John Muir's "bee pastures," the carpets of gold and purple and red flowers of the hillsides and of the great interior valley to the north.

Some of the worst forest fires in the West occur

245

in the mountain chaparral after it has dried out in the summer. Because of the denseness of the shrubs, it is almost impossible to control these fires, which spread rapidly over thousands of acres, sending up a yellow-gray pall that obscures the sky. Fires, however, are something that the chaparral plants have learned to live with; many of the species have large root crowns at ground level that survive the blaze and sprout new stems with the first rains. Others have seeds that germinate much more rapidly after being subjected to heat.

Canyon streams dry up in summer, as do streams in the Mojave. Along the banks of the former can be found many species of desert wildlife. The roadrunner is here, under a different name—the chaparral cock. Hummingbirds dart among the wildflowers, nearly always seeking nectar from the bright red ones. And the ringtail, an animal the size of a house cat, prowls under wild blackberry and snowberry and such canyon trees as alder, sycamore, poplar and bigleaf maple. The pack rat, white-footed mouse and other small rodents are the objects of its midnight searchings.

With so many berry bushes and rodents, this countryside was once the haven of three kinds of grizzly bears, exerting their hump-backed supremacy to exclude the black bears. Never caring to climb, as their smaller relatives do, the grizzlies lived almost entirely in the chaparral until they were killed off by man with his dogs, traps, poisons and .30-.30 rifles. Today only a few survive, in the Sierra del Nido south of the Mexican border.

The most unique of southern California's wild creatures is the California condor, whose great wings, spanning nine and a half feet, may not cast shadows across the canyons for many years more. The story of the bird's decline closely parallels the state's phenomenal population growth and industrial development since World War II.

Writing in 1894, John Muir foresaw that the shepherds "with their flocks of hoofed locusts" would trample the perennial bunch grasses of the Central Valley, leaving only the slender wild oats, the filaree and bur clover, annuals brought in as seeds in the wool of Spanish sheep a hundred years before. He knew that a time would come when the land "will be tilled like a garden, when the fertilizing waters of the mountains, now flowing to the sea, will be distributed to every acre, giving rise to prosperous towns, wealth, arts . . ."

What he could not have imagined is the extent of California's population explosion. In 1940 there were 6,907,387, and by the end of 1964 there were 18,500,000 people, with immigration swelling the number at the rate of 1,000 per day.

The chaparral-covered hillsides around Los Angeles have been bulldozed and terraced for housing, inviting erosion and flooding. Southeast of San Francisco, the Santa Clara Valley has lost more than 80,000 acres of prime vegetable farms and orchards to defense factories, shopping centers and tract houses. Megalopolis extends all the way from the Golden Gate south to San Diego, and in some places inland half way across the state. Developments of desert shacks and motels with swimming pools press right up to and beyond the old established Hollywood retreat of Palm Springs, under the snowy peak of San Gorgonio. Names like Apple Valley belie the creosote bush that in many places stretches unrelieved to the horizon.

The water for the crowded south, especially the booming desert, comes from the Colorado-River dams to the east. But Arizona has won its claim to a large share of this water, and California must now utilize northern rivers that spill into the Pacific Ocean. Ten dams have been built on the Sacramento River since 1938, and the first of twenty planned for the Feather River was dedicated by Governor Ronald Reagan on May 5, 1968.

When all this began, the condor was a lonely relic of the Ice Age, the species only about sixty individuals strong. The state has protected the birds by law since their Sisquoc Falls roost in the Los Padres National Forest was set aside as a refuge in 1937. By 1947, the 53,000-acre Sespe Refuge —also in the Los Padres National Forest—had been established. While the human population was undergoing a vast increase, the condors' numbers dropped by about one third, according to observations made in 1964. No longer were large groups of condors seen feeding together, though carcasses of cattle, sheep and deer were plentiful. The big scavengers were simply frightened away by people. They did not fly off very quickly at the approach of a man; but once in the air, an entire flock would not settle down to roost or feed for half a day after being disturbed. If their nests in the rugged Sespe canyons were discovered, the

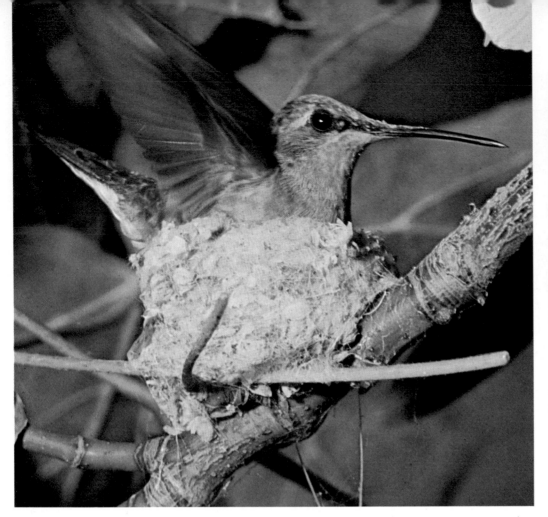

The monkeyflower has convenient openings to probe.

With wings that never seem to stop beating, a female Costa's hummingbird tends her nest.

A Nesting Hummer

To fuel its tiny, energetic body, the hummingbird that nests in a Western canyon must eat half its weight in nectar daily. With a specially adapted beak, it probes the corollas of a large variety of flowers—preferably red ones, such as shown here.

One favorite of the hummingbird is scarlet honeysuckle.

A columbine is inviting.

247

parent condors often abandoned their eggs or young.

This kind of thing can be disastrous for a slow-breeding bird. The condor needs five years to mature. Its offspring remain flightless for months and dependent for nearly two years. Adults cannot nest more than once every other year, and each time they fail, the population dwindles. Consequently, every effort has been made to prevent people from entering the condor sanctuary—and much ill feeling has resulted. Townspeople resent having an area completely withdrawn from possible development.

Deer hunters believe that their activities in killing game and leaving parts of the carcasses behind help to feed condors. But the condors are probably alive today because the National Audubon Society and a few local ranchers have kept a close watch on the Sespe and the upper drainage of the Sisquoc Canyon. Observers stationed at various strategic points, such as Reyes Peak, have sighted about 53 condors, and the proportion of young birds indicates that the population is stable or else increasing slightly. Being shot at and feeding on "1080"-poisoned rodents seem to be greater hazards than failure to nest.

How long the region can be kept wild in the face of encroaching megalopolis, however, is another question. Los Angeles is already spreading into Santa Clara Valley, where suburban housing springs up on former feed lots. Too much pumping of underground water has caused an incursion of sea water under more than 8,600 acres of the Oxnard Plain. To solve its water problems Ventura County looks to the Sespe Creek, in the heart of the condor nesting area. So far, the ten-year fight of the water interests to build the Topatopa and Cold Spring dams has not been won, and special opposition has built up because of the fact that an access road would have to run through the heart of the refuge to Topatopa Dam from the town of Fillmore.

The fate of the condor hangs on politics, because water districts are established on economic and political, not ecological, lines. Knowledgeable people in the area point out that the proposed dams are too far upstream to stop most of the floodwater that comes down after a heavy rain. The major drainages of the West Fork, Little Sespe, Maple Creek, Tar Creek and Alder Creek are all below

the dams. Furthermore, flooding occurs in only about one year out of four. Most of the water that comes from the Sespe doesn't run off, but goes into the underground supply which is pumped out by ranches farther down the valley.

The proposed $90-million Sespe Creek Project would provide 27,000 acre-feet of water—on paper; 15,000 more would come from the Southern California Metropolitan Water District, and 20,000 from the state (perhaps from the new Oroville Dam on the Feather River). Authorities feel that this would be adequate until 1990. Then what? Desalinization of sea water? And if so, why not now? Perhaps the persistence of an Ice Age vulture will prevent Californians from making the costly mistake of depending too much on the watershed potential of the coastal ranges. Without the condor's ungainly presence, the alternatives might not be explored.

The high Mexican tablelands that enter West Texas and southern New Mexico are similar in extent, altitude, rainfall and number of frosty winter days to the sagebrush country of the Great Basin. A climb up rugged Lost Mine Trail in the Chisos Mountains reveals a sampling of the varied plants and wildlife of the Mexican ranges tumbling away to the south in a vista that extends for so many miles that—in the local phrase—you can see "the day after tomorrow." Like Lost Mine Peak, elevation 7,650 feet, none of the Mexican desert peaks in this panorama are high enough to sustain more than a scattering of the tall pines and fir characteristic of the bordering Sierra Madre.

Where Lost Mine Trail ends, piñon pine frames the view of the South Rim. Elephant Tusk and other peaks loom in front of the Big Bend of the Rio Grande, about four miles away as the eagle flies. Below lies mixed woodland of pine and oak. Many varieties of Mexican oak grow here. They are very difficult to identify because of the variations in size and shape of the leaves, even leaves on the same tree.

As one descends from Lost Mine Peak, at

A Mountainous Desert
Seen from Lost Mine Peak, the ranges of the Chihuahuan Desert fade southward into Mexico. Nearly half of this mountainous wasteland is over 4,000 feet high.

The Hog-nosed Skunk

A naked, piglike snout distinguishes the rarest and most nocturnal skunk north of the Mexican border. It is difficult to trap with ordinary skunk bait, often ignoring the lure but rooting up earth all around the trap in search of ground beetles instead.

The Mexican Jay

Perched on an old tree near a picnic table, a Mexican jay looks for a handout. This species has unique communal habits; as many as four birds may build a single nest, place all their eggs in it, and share feeding duties after the young jays hatch.

The Gray Fox

Slinking across stones, a gray fox follows the trail of a small animal in the brush. Mice and rabbits make up the bulk of its diet, but this carnivore also savors an occasional meal of insects, berries, and the young or the eggs of birds.

about 5,000 feet or below (depending on whether the slope faces south or north) grama grasses and mesquite chaparral begin to appear, and spread downward onto the plains. Sometimes the mesquite is replaced by sotol, raising on a single stem a narrow inflorescence of white blossoms, by agaves, or by spectacular giant daggers and other yuccas. Together with dozens of kinds of succulent cacti, these plants grow in the rockier, better-drained places. On the lowest, driest flats, the ubiquitous creosote is relieved by an occasional mesquite, allthorn, or flaming ocotillo along a wash.

Most of the birds make their nests, the carnivores their dens, and the white-tailed deer their pastures in the piñon-juniper and oak woodlands. Only one jay—the pale blue, crestless Mexican species—inhabits the Chisos, chattering noisily as it searches the trees for acorns, seeds and in-

sects. At campgrounds, you can observe how closely the jays follow a daily routine. In groups of a dozen or more, they comb the hillsides for scraps of food. A flock quickly converges on the spot where a scout announces that he has found something edible. Another reason for rallying is the discovery of a predator—hawk, owl, fox, or bobcat—which the jays customarily mob.

If you see a shadowy form smaller than a coyote exploring the bushy draws during late afternoon or early morning, it is probably a gray fox, more common in the Southwest than anywhere else. The animal is at home among trees, and climbs the trunks with ease. This is not just a means of escape from dogs. More than once a gray fox has been surprised while sunning itself comfortably on a broad limb. (Tropical woodland foxes in Mexico, with sharper and more curved claws, are even better equipped for climbing.)

The hog-nosed skunk sleeps the day out here in a rock pile or hollow log. The only skunk found also in South America, it has extremely coarse fur and a long, naked, pig-like snout that it uses to root up big patches of soil in pursuit of beetles and other insects. This tropical skunk is not as numerous or as prolific as its northern relatives.

Ungulates ranging in size from the collared peccary to the pronghorn antelope and the mule deer may be seen on the mesquite grassland, together with jack rabbits, pack rats and ground squirrels. With the spread of the mesquite in recent years, the diet of all these animals has become principally mesquite beans. Just as the creatures of the Great Basin depend upon sagebrush, the wildlife here makes use of the most abundant plant of this habitat.

Although the antelope shun the low desert, mule deer enter the dry breaks along the Rio Grande to cut the stalks of agave and paw open the cabbage-like heads of sotol. Peccaries, the only wild pigs native to the Western Hemisphere, shelter in the thickets on the higher slopes, and spread out over the desert to feed. In West Texas, the peccaries feed mostly on prickly pear, which provides both food and water.

Pushing up bundles of succulent, stiletto-shaped leaves to a height of eight or ten feet, giant daggers completely fill one shallow valley about eight miles long in the northeastern corner of Big Bend National Park. The white, bell-like blossoms come out in March and April, a month before the agaves and two or three months before the cactus plants flower in the summer rainy season. Rodents live in the dry leaves at the dagger's base, and its crown may hold the nest of a white-tailed hawk. This Southwestern species, lighter in color than the red-tailed hawk, ranges no farther into the United States than the giant dagger country.

This is the country of the largest of all spiders, the tarantula. The powerful, predaceous creature, armed with strong fangs and potent venom, will kill and eat any mouse, lizard, small snake or insect that it can find on the ground. It predigests

Dagger Flat
*Giant daggers, their trunks cloaked in old leaves,
dominate a mountain basin in Big Bend National Park—
the only large stand of these plants in the U.S.*

its food by flooding a wounded victim with secretions that soften the tissue and allow it to be sucked into the spider's body. Thus the killer can reduce the bulk of a grasshopper in a few minutes, of a small rattlesnake in a matter of hours.

But tarantulas are neither belligerent nor particularly dangerous to people. And they have many problems of their own. Poisonous hairs on the body fail to provide complete protection for the males, which are often eaten by their mates. Both sexes are plagued by numerous enemies, including parasitic flies and wasps. Since more females survive than males, it is usually the female tarantula that falls prey to the tarantula wasp (a digger wasp). She is stung, carried off to the lair of the wasp, and left paralyzed in its sealed burrow to provide fresh food for the wasp's larvae.

Ever since Big Bend National Park was established in 1944, owners of sheep and goat ranches bordering the park have been convinced that predators move in from the protected area to prey on their livestock. A typical comment runs as follows: "Eagles and panthers and bobcats are breeding all over those mountains, and the government doesn't do anything about it. Here we are paying for predator control and the government is raising the varmints up there in the hills."

This attitude does not take into account the laws of animal territory, possible increase and migration. Or the fact that outside the park there are similar mountains in which not all coyotes, bobcats, panthers, and eagles could possibly have been exterminated, even though predator control has been practiced since 1890. It is just not economically feasible to search out every den and nest in these rough desert areas. Meanwhile, the predators that survive become more destructive and wilier.

Especially hated is the golden eagle. But how many eagles could the park possibly shelter? If each pair needs 25 square miles—as is the case along the Front Range of Colorado—there would be enough territories in Big Bend for only about ten pairs. If each pair raised one young bird every year, there would be an annual crop of ten eagles to spread out over 30,000 square miles to the north and east, without considering Mexico. As a matter of record, only three pairs were breeding in the park some ten years ago. It is not certain

A Tarantula at Table

Its fangs imbedded in a grasshopper, a tarantula drains its
prey of all body fluids. This arachnid is an ancestral type of
spider. It lives longer—sometimes 25 years—and grows up
to three or four times the size of the true, or modern, spiders.

255

that any at all remain there now. In the pre-dawn mist of the Chisos Mountain Basin, we saw a pair of golden eagles circling overhead two summers ago. We could not know whether they had a nest or young that year or, if they had, how long the birds would survive. The people who have to cope with the problems of tending livestock in West Texas—ranchers and members of sheep-shearing crews—do not hesitate to shoot eagles in their precarious sanctuary.

Eagles have been shot from planes in the Southwest since the early '30s, but the coyote drew most of the attention of five aviation clubs in Texas and New Mexico until the decade between 1940 and 1950. By then the use of the new poison, "1080," had drastically cut the number of coyotes. So pilots intensified their efforts to clear ranches of all the eagles they could find a month or two before most of the lambs were born in March and April. Men would take to the air and pursue the big birds with a 12-gauge shotgun mounted on the fuselage. William Hargus, of Fort Stockton, a former fighter pilot, tells of the tactics necessary to shoot one down. If Hargus flew alongside an eagle going 40 miles an hour, with the plane moving at 50, there would be a good chance of emptying a charge into its body. If the split-second opportunity passed, however, he had to follow the bird as it circled upward on a thermal, then overtake it and shoot it down from above. Apparently quite a number of eagles have been shot at and missed. These birds quickly learned to drop to ground level, where it is difficult if not impossible to shoot them from a plane.

A Grisly Warning
In West Texas, the remains of coyotes and bobcats line the gate of a member of the Texas Sheep and Goat Raisers Association, which wars on all such predators.

A Prime Scapegoat
The golden eagle, a bird that lives in West Texas and also migrates there from northwestern North America, is blamed for much of the mortality of lambs and kids from all causes in the Marfa and Davis Mountains region. After the poisoning and shooting of eagles for thirty years, conservationists estimate that fewer than 10,000 survive on the entire continent—and the Texans still have problems.

Nevertheless, no method of killing eagles has been more effective than aerial pursuit. The Bureau of Sport Fisheries and Wildlife reports only 60 eagles taken from the Davis Mountain Foothills during the entire winter of 1940 by poisoning, compared to 23 eagles shot during six hours of flying time in the same area. The largest number shot down by William Hargus in one day was 38, a record for some years. After combining all the claims and making allowances for "the one that got away," it seems likely that about a thousand eagles were killed annually over a period of 20 years.

Conservationists finally realized that the vast majority of the eagles being shot in the Southwest were not residents of that area, but were migrants from Canada, Colorado, Montana, Wyoming and Idaho. The great reservoir of predators was not "up in the hills," but up in the North. On the currents of the Texas "northers" each fall, these travelers filled up the canyons and soared out over the flats and valleys of the Trans-Pecos.

Shooting these birds depletes the relatively small number of eagles—about eight to ten thousand—that inhabit the entire North American continent. Made aware of the situation by conservationists, Congress in 1962 passed the Golden Eagle Law, making it illegal to shoot eagles from planes—although killing by other means is still allowed if the birds are doing damage to property.

This does not mean that the ranchers have given up their plane clubs, and a few pilots are said to be still hunting eagles in defiance of the law. There is a general belief among ranchers that

257

Land of the Creosote
Nothing but creosote bush occupies sandy flats in Big Bend National Park.
This bleak landscape was long ago over-grazed by cattle and eroded
by flash floods and winds of the Chihuahuan Desert. No one knows when, if ever,
the desert grasses that may have existed here once will grow again.

predation has increased. Near Alpine, Texas, and in the rolling foothills of the Davis Mountains, sheep and goat raisers feel certain that most of their losses of lambs and kids are due to eagles. In 1962, these losses ranged in the hundreds for herds of thousands. Each animal killed represented a loss of 12 to 15 dollars. It should be remembered, however, that many lambs die of starvation and sickness on the range, and any coyote or eagle that feeds on the carcass may be blamed. Consequently, the estimates of killings by predators are considerably exaggerated.

To help solve the problem, the Bureau of Sport Fisheries and Wildlife has been carrying on a long-term survey of eagle habits. In the past two years, researchers have been clipping patterns in the wing and tail feathers of young Northern birds while they are still in the eyrie. After they have migrated in the fall to the warmer ranges of the Southwest, other researchers will be able to tell exactly where they came from. Still in the experimental stage is electronic tracking by miniature radio sending devices attached to the legs of young eagles. These are tiny broadcasting sets similar to the ones worn on collars by grizzly bears in Yellowstone National Park that enable scientists to track every movement of the bears.

For the continent as a whole, the conservationists are surely correct in stating that rabbits and rodents comprise the major food supply of golden eagles. The big birds are often seen along roadsides in New Mexico and West Texas feeding on rabbits, killed either by them or by passing cars. And studies of nesting birds in the region have shown that rabbits are the staple of both adults and young during the spring. Yet large groups of immature eagles do gather in the lambing pastures, and there is no doubt that they kill some lambs.

On the whole matter of damage done by eagles there are many opinions, but few facts with which to make a quantitative judgment. The rancher sees five or six eagles in his lambing pasture. During his lifetime he has witnessed perhaps two or three kills of lambs. He has very likely seen eagles eating lambs that died from other causes. Whatever the truth may be, he *believes* that eagles are responsible for a large proportion of his losses.

On the other side of the question, there are still some conservationists who would like to believe that eagles kill no lambs. At present, the only support they can find for this view is theory. By calculating the number of pounds of rabbit meat produced by a pair of rabbits each year in the form of young rabbits, and multiplying by the number of adult animals per square mile, they show that the rabbits provide enough food for an eagle ranging over the territory. This theorizing passes over the natural inclination of hawks, eagles and other predators to feed on whatever is easiest to catch, not necessarily the most abundant prey. Furthermore, individual birds can develop special food preferences. Some Idaho eagles feed their young on marmots or grouse instead of rabbits.

The case against the eagles seemed to be clinched when biologists recently found lamb bones around an eyrie in New Mexico. But this was a unique pair, greatly different in habit from the majority of eagles under observation throughout the Western states. It is not reasonable to conclude that all golden eagles should be shot for the sheep-killing activities of a few.

Analysis of the West Texas problem is complicated by the fact that numbers of juvenile eagles stay on their wintering ranges until early April. Too young to breed, these birds have no reason to hurry north. (Courtship among the adults begins in January or February in Montana and Idaho.) Do any of the young migrants stay in Texas and replace nesting pairs that have been shot? Until more has been learned about the movements and other activities of eagles, there is little reason to believe that we can either "control" them or "save" them.

At the base of the eagle problem in West Texas is the land itself. A century ago, the mountains and basins supported vast herds of wintering bison and comparable numbers of pronghorn antelope. By 1900, only remnants of the former wildlife populations remained, and the grasslands were overgrazed. By the time Big Bend National Park was established, no one could remember whether the desert flats were once grassy or had always been barren. Here, as in Arizona, there had been drought and arroyo-cutting, and cattle herds that destroyed the plant cover. Bunch-grass flats thrived when there was periodic flooding. But with the delicate balance between the land and the water upset, there are now only small fragments of tobosa grass left on Tornillo and Tobosa Flats in the national park. Ungrazed except by wild animals since 1944, these two areas are little outdoor laboratories in which we can study the recovery of an environment. In the desert, such recovery is slow. Under the long-term drying trend of the Southwestern climate, the bottom lands may have reverted to desert for our lifetime or longer. If this is so, West Texas faces troubles more serious than predators.

Instead of retiring depleted cattle ranges, the livestock industry has found it more profitable to use them for sheep and goats. The sheep population during the last 30 years has grown from a few thousand to over six million head, the goat population to nearly four million. To encourage the renewal of grass, much woodland and chaparral has been uprooted or killed out with herbicides. This has had the result of exposing newborn sheep and goats, as well as antelope kids and the fawns of deer, to attack by eagles.

At one time there were no recorded sightings of eagles in Kerr County, Texas. By 1946 the eagle was an occasional winter visitor. In recent years, 25 to 30 have been killed annually in this ranch country. Something must be causing the birds to concentrate in places where they have never been before; perhaps it is the combination of more livestock and less cover. On the stony land, with its sparse vegetation, the sheep must spread out to feed alone, making them particularly vulnerable to predators.

Since the desert has encroached upon thirty to forty per cent of the former grass flats in Texas, mule deer and antelope range farther up on the hillsides, and the eagles acquire even more enemies by preying on the wild kids and fawns. Ranchers have a monetary interest in keeping game animals on their land. Whatever competition such animals give to the livestock in arid pastures is compensated for by the sale of private hunting permits. The number of these permits allowed is based on the game census. An antelope permit sells for $40, a deer permit $100. With deer and antelope populations increasing, the permits may yield the sheep rancher several thousand dollars in a good year, to add to the return on his wool and lamb harvest. Thus the rancher gets just as angry about the killing of a fawn on his property as about the loss of a lamb.

In dusty West Texas, little is being done to improve the condition of the land. Both wildlife and domestic animals tend to cluster in certain wide, lush basins of the Valentine Plain, the Marfa Plateau, and the Davis Mountains. Elsewhere, with the rivers and springs dried up, life centers around ponds and reservoirs supplied with water from underground sources by windmill and gasoline pumps. Even in wetter years, the amount of usable land is very small compared with the wastelands. Each year the desert encroaches more.

North of the Mexican highlands and the Big Bend of the Rio Grande River, lava flows and white sand dunes extend the Chihuahuan Desert into the basins of New Mexico. The dunes in the White Sands National Monument, near Alamogordo, are composed of gypsum fragments. In a more humid region, the particles would be carried away in solution by surface and subsurface water. But here the gypsum crystals are washed down off the mountainsides into shallow marshes that evaporate during the summer. The prevailing southwest wind then churns up the sand grains, and whirls them away to the dunes. The glistening white hills, covering an area of nearly 500 square miles, are unequalled in the United States—if not in the entire world—in their extent and purity. As the sand is driven up a rippled slope to the crest of a dune, it buries the little yucca or soap weed growing there. But often the plant manages to struggle to the surface. In some places, the migratory dunes have left behind pedestals of compacted gypsum nearly 40 feet high, bound together by yucca roots.

Among the white dunes, dark-colored animals are easily seen and captured by predators. So the only lizards, spiders and mice that have survived and reproduced over the ages are pale ones. A little pocket mouse is straw-colored in the dunes, though in the nearby red hills individuals of the

White Sands
Patterns in the gypsum sand dunes of New Mexico exist for a moment in desert history. Pushed by persistent winds, the crest will move on across the Tularosa Basin.

same species are rust-colored, and in the beds of black lava they are black. Wherever rocks of different colors mingle, the populations also mix, and both rusty and black mice run back and forth over the mottled surface.

A reminder that some of our deserts were once tropical forests is the volcanic plain of the Painted Desert in northeastern Arizona. Here cap rock of sandstone and lava holds together layers of bentonite, stained with the red, blue, brown and yellow of manganese and iron oxides. Under gathering summer rain clouds, these buttes and mesas glow with an intensity seen in no other badlands. Erosion, the force that made the desert, continues, as seasonal gully-washers lay bare 200-million-year-old fossil tree trunks. It was once thought that the organic matter in this wood was replaced, molecule for molecule, by mineral matter. We now realize that the network of woody tissues became impregnated with solid substance before it decayed. The continuing presence of organic molecules in the petrified wood accounts for its brilliance and variety of color.

The periods of great mountain-building in the West raised these tropical flood plains, on which grew trees related to existing pines in Australia and South America. In the aridity of the new climate, the mesas are crumbling, and ancient logs tumble out of the canyons and ravines. Each crack in the petrified wood may record a quake on the restless face of the earth.

7
New Mountains, Old Plateaus

As mountains rise and volcanoes erupt, the earth takes on a new face, to be gradually aged by the forces of wind, water and ice. The landscape of the American West is still craggy from the effects of earth movements in our own time and in the recent geologic past. Only tens of millions of years ago the Rockies were thrust up, either in complicated folds as in the Lewis Mountains in Glacier National Park, or as great tilted fault blocks like the Tetons in Wyoming, which are believed to be still rising. Even newer mountains are the Olympics and the North Cascades in Washington. The Coast Range and Sierras of California and the basin ranges of Nevada are part of an unusually active land belt surrounding the Pacific Ocean, where most of the total earthquake energy of the earth is released.

The age of the last uplift in these areas is figured in tens of thousands of years. But the volcanoes that overshadow our Western landscape have remained active until quite recently. The last major outburst of Mount Rainier occurred only 500 to 600 years ago, according to radiocarbon dating of tree stumps buried in ash. Still younger are the thin sheets of powdery white ash (composed entirely of tiny fresh glass shards) that spread in a blanket through central and eastern Washington, and probably came from Mount St. Helens. Though dormant at present, the great volcanoes of the coastal region are not dead. Wisps of steam issue from their craters, and one —Mount Lassen—has erupted several times in this century.

Perhaps no kind of mountain is more subject to erosion than the perfectly-formed volcanic cone. For this reason there is probably no such cone more than a million years old. Mount Rainier's crater was reduced by explosions, rock falls, glacial erosion and slides from a former height of 16,000 feet to about 14,000 feet before the end of the last Pleistocene glaciation. (We know this because of the inclination of lava and cinder layers on the mountain's flanks, which are indicators of the old slopes.) Great valley glaciers left gigantic amphitheaters, or cirques, separated by thin ridges that are called aretes. Then, some 5,000 years ago, the glaciers disappeared. Melt waters stored in the porous volcanic rocks of the mountain supplied streams in the valley heads all around the cone—and these streams frequently flooded, devastating the slopes. In a period of extremely cold weather, sometimes called the "Little Ice Age," about 4,000 years ago, the present small glaciers of Mount Rainier were born. They have been waxing and waning with the changing climate ever since, and the erosion continues.

The Rockies once towered as high as the Himalayas. Today, after millions of years of erosion,

Rocky Mountain peaks still project through the clouds at altitudes of more than 13,000 feet, like islands in a billowing sea, and are mantled with almost eternal snows. From these snow fields come the rivers that have cut the deep canyons of the Gunnison and the Royal Gorge in Colorado and the Middle Fork of the Salmon River in Idaho. The high plateaus to the west, comprising approximately 15,000 square miles of Arizona, New Mexico, Utah and Colorado, have been eroded on a scale unmatched in any other part of the country. From the wide, open valleys and receding cliffs of this arid region, perhaps 10,000 feet of strata have been washed away, mostly by the Colorado River, its tributaries, and their tributary streams—the makers of the canyonlands.

From ten to one million years ago, the plateaus rose up, while at the same time the old, winding rivers crossing them were cutting down into the rocks. The result is a pattern of meanders and steep-walled canyons, especially where the land fractured and gently domed. Each of the plateaus north of the Grand Canyon of Arizona—from west to east, the Shivwits, the Uinkaret, the Kanab and the Kaibab—represents a 1,000-foot increase in altitude due to this fracturing. The Kaibab (which literally means "mountain lying down" in the language of the Paiutes) rests at 9,000 feet above sea level.

The Colorado River and the Grand Canyon describe a great semicircle around the Kaibab Plateau in the eastern part of Grand Canyon National Park, and a smaller curve around Havasupai Point; then the river loops around Great Thumb and Chikapanagi Points, a distance of 217 miles, before leaving the park on the west. It is at the Kaibab, the highest point, that the Colorado River has cut deepest. Here erosion by tributary streams, snow melt and landslides has widened the canyon most, creating the best examples of "its colossal buttes, its wealth of ornamentation, the splendor of its colors, and its wonderful atmosphere," as described in a geological survey report by C. E. Dutton in 1882.

It is fascinating to discover the contrast in climate zones between the South Rim and the North Rim of the Grand Canyon, separated by nine miles of chasm (on the average) and 2,000 feet of altitude. On the South Rim grows piñon pine; on the North Rim, aspen, bent into weird shapes by the heavy winter snows. North and east of the Grand Canyon, the heights at Zion Canyon are covered with piñon pine and gnarled juniper; Bryce Canyon, at 8,000 feet, reaches the ponderosa pine belt; and Cedar Breaks, at Brian Head, becomes almost arctic-alpine at 11,315 feet.

On the high plateaus and in the mountains you can travel a few miles up or down in one

area and encounter climate and vegetation equivalent to all the latitudinal zones except the tropical. To find all of them on a single mountain, however, is rare. The mountains from New Mexico southward are extremely varied in climate because they rise from hot deserts, but at that latitude the peaks are not high enough to have tundra at the tops. Tundra begins to appear at 11,200 feet in the ranges of Southern Colorado; northward, in Montana, it starts at 8,000 feet. In the Pacific Northwest, where snowfall is heavier and perennial glaciers cap the mountains, the treeless tundra extends much lower—growing as low as 6,000 feet in the Olympics.

The pattern of plant zones on a mountain varies according to the exposure of the slope—north or south—and also according to the soil and the amount of moisture available. Tongues of conifers and aspens push down from the heights into deep, cool canyons along streams. Sagebrush sweeps from the bottom to the very top of a dry mountain that may support firs on another, wetter side. In these situations, animals live at altitudes where they are not expected to be—the Colorado chipmunk in the canyon, and the scrub jay on the mountain top.

To hike up a mountain trail is to cross countless times a clear, bubbling stream that has the chill of melting snowfields and glaciers. Cold waters like this rush downward from their unseen sources through the valleys, nurturing hillside flower gardens all summer long. The peak of the first floral display comes in July, with white avalanche lilies and yellow glacier lilies poking through the snowbanks. About a month later, the same nooks and crannies will be filled with the bloom of Indian paintbrush, penstemon, monkeyflowers and other species.

The streamside gardens are mere ribbons of growth, never spreading farther than the trickles of water that find their way through coarse soil and boulders. The "big meadows" are mostly bottoms of old glacial lakes, gradually filled by landslides with just enough fine soil for the growth of sod and tall-stemmed flowers. The most celebrated of these meadows flourish where the snow pack is greatest—above Yosemite Valley in the Sierras; on Hurricane Ridge in the Olympics; and in the Paradise and Sunrise Valleys on Mount Rainier. Snow depths of 150 inches (with a water content of 50 inches) are not uncommon in these areas. The soil may not freeze all winter under this insulating layer of snow; if it does freeze before the first snowfall, the inner heat of the earth and the energy of the sun combine to thaw it quickly in the spring. So the meadow soaks up an ample supply of water before runoff occurs. Surrounded by forests of alpine fir, an open field of grass and flowers may exist for centuries, but sooner or later enough soil will wash down from above to make a bed deep enough and dry enough for gilias and composites, and trees may take root.

Though many of the same flowers grow in the Olympics, the Cascades and the Rockies, the forests in each range are very different. Every Northwestern mountain range acts as a barrier to Pacific rainfall, and this limits the eastward spread of trees from the Pacific forests. Thus, the Sitka spruce and bigleaf maple of the rain forest are not found east of the Cascades. Other trees, such as the Douglas fir and lodgepole pine, which may grow up to four feet in diameter in the Cascade Mountains, dwindle to about a foot in diameter as rainfall drops off near the Continental Divide.

Along this imaginary line—a line that follows a series of high peaks and ridges in the Rockies—any precipitation that falls on the west slope runs into the Pacific Ocean or the Gulf of California; any that falls on the east flows into the Mississippi River and the Gulf of Mexico.

Here the great river systems rise. On the east slope are the sources of the Missouri, the Platte and the Arkansas, main tributaries of the Mississippi, and the Rio Grande, which flows from the San Juan Mountains in southwestern Colorado directly into the Gulf of Mexico. On the west slope, waters from the Colorado and its tributary, the Green River, flow to the Gulf of California; and both the Clark Ford-Pend Oreille river system of Montana and the Snake River, which begins in Wyoming, join the mighty Columbia on its way from the Canadian Rockies to the Pacific.

A High Mountain Stream
At an altitude of about 9,000 feet in Glacier National Park, Montana, melt waters from Sperry Glacier plunge over the steep rock wall of a hanging valley.

A sizable proportion of the water that feeds these rivers comes from snow, locked in the heights all winter and then suddenly released in the heavy seasonal flow of late spring and early summer melt-off. For more than thirty years past, snow surveyors on snowshoes and skis and in ski-equipped airplanes have been measuring the depth of high-mountain snow, calculating its water equivalent, and very accurately predicting flood or drought from year to year. They do it by sampling a water course from high to low elevations, taking an average of the readings, and then adjusting for soil conditions and rainfall.

In some places the readings are now taken by electronics. Solar cells powered by the sun on a high mountain slope charge batteries that can operate at temperatures of 60 degrees below zero. The battery-powered equipment relays information about the depth and water content of the snow from a "pressure pillow" to a weather station

The Great Divide
*Seen from Sperry Glacier, rugged
peaks of the Continental Divide sweep
northward into Canada. All the water
flowing from this western slope
empties into the Gulf of California
and the Pacific Ocean. During summer
melt-off, the snow that has filled
the glacial crevasses disappears, and
sometimes streams of water open
gaping cracks to widths of about twenty
feet and depths of more than 200 feet.*

every time a button is pressed. The pillow is a pancake-shaped rubber bag filled with methyl alcohol. It is about 12 feet in diameter and six inches thick. When snow falls on the bag, a change in pressure occurs; this change is converted into inches of water on a manometer—or, in the more recent work, into electrical units of measurement that can be sent to a base station by radio telemetry.

On predictions of snow melt-off, a serious water shortage was averted by building reservoirs and taking other conservation measures in Utah in the year 1934. A flood on the Columbia River was greatly minimized in 1950. But there are many water problems that have not been attacked until very recently.

In 1961, the flows through the reservoirs along the Snake River in Idaho were so low that the oxygen in the water was used up by organic pollutants, and 250,000 fish died. The dead fish added to the pollution, and the city of Twin Falls had to drill new wells to obtain water. A different pollution problem, caused by flooding, is sedimentation. When the flow of a river is above normal, the soil carried down from the mountains spreads out and chokes up the aquifers of the flood plain. Contoured hillsides, strip cropping, terraces and grassed waterways are all no more than partial solutions. They help save the land, and they prolong the lives of the big reservoirs built to hold surplus water. But we still have floods.

Some years ago the Swiss Institute for Snow and Avalanche Research started an experiment that may result—both in the Swiss Alps and in our own Rockies—in some control of water flow at its source, on the windswept ridges of the high mountains. Giant snow fences, the Institute discovered, could be made to cause drifting and prevent avalanches. The pattern of natural drifting was simply altered so as to pile up snow in places where a slide would be unlikely. As a by-product, the drifts formed huge water banks at an elevation where a steady rate of melt would feed the streams and rivers below at a controlled pace all summer long.

Consideration is being given to the erection of such structures on eighty avalanche paths in the Colorado Rockies. One snow fence already stands at an altitude of more than 12,000 feet on a ridge above Independence Pass, near Aspen, its sturdy pole construction rising high to catch the drifting snow before it goes over the steep rim where it might become an avalanche. In this area the snow clouds begin to race across the peaks in late July, bringing flurries two or three times a day, harbingers of the long tundra winter. If the fence holds, it will conserve the highways, the land, and the supply of water for a major part of the Colorado River drainage.

Crevasses are the signs of life in a glacier. In the heat of a summer day, the river of ice expands and explodes noisily, its cracks swallowing up avalanches of surface snow and rocks. The glaciers of Glacier National Park are small and relatively inactive now, but the surrounding scenery is a grand display of the work of old valley glaciers that existed in at least two geologic periods, each lasting thousands of years, during the Pleistocene Epoch. Sperry Glacier, 330 acres in expanse and the biggest of the glaciers now in the park, occupies a cirque that was cut away by an ancient glacier below—a feature of the glaciated landscape that is called a hanging valley. Of course, the great ice flows that cut gigantic stairsteps in the mountains have long since gone. The new glacier is so puny that it cannot cover the deep scratches in the rocky floor of the valley, some of which are now lakes filled with melt water.

Old glaciers fashioned the series of rugged ridges and peaks of the Continental Divide, on the northern horizon—quarrying out innumerable sharp horns, as in the Swiss Alps. Where ice from two valleys once nearly converged, the result is "The Garden Wall," perhaps the most famous of the aretes in the northern Rockies.

To see large "working" glaciers anywhere below the Canadian border today, you must go to the Pacific Northwest, where the snow line is low and icefalls on the steep slopes are cracking and actively flowing. Five separate streams of ice make up Blue Glacier, high and difficult of access, in the Olympic Mountains of Washington. Considerably below the permanent snow line on Blue Glacier is a 1,000-foot icefall that, by its speed and compression of ice and sediments, produces complicated patterns similar to those of the Alaskan valley glaciers. From a distance, its vertical foliation looks like a series of arcs and its horizontal ogives appear as alternating light and dark bands.

Our volcano glaciers are less complex, but easier to reach and just as interesting to observe. Mount Rainier alone carries 40 giants on its slopes, a total ice area of 34 square miles. Nisqually Glacier descends to an altitude of about 4,500 feet. Both the U-shaped valley and the terminus of the glacier can be seen from a bridge that crosses the Nisqually River. Along the sides of the valley are rocky ridges—moraines that mark the maximum extent of modern glaciation, around 1840. From the Moraine Trail, along the eastern ridge, hikers look down into a grumbling

A Valley Glacier
Churning up volcanic lava and ash, Nisqually Glacier, a tongue of ice hundreds of feet high, descends the precipitous south slope of Mount Rainier.

mass of ice, mud, and rocks. Fed by the summer runoff from the steep south face of Mount Rainier, a river pours from the tongue of the glacier.

Until 1946, even this noisy glacier was rapidly shrinking—losing more water from melt-off than it could gain from snowfall on the mountain top. Then a U.S. Geological Survey team reported that the ice on its upper reaches was becoming thicker and that new tongues were beginning to push forward. By 1964, Nisqually was advancing at the surprising rate of 100 feet a year. Most other Mount Rainier glaciers thickened, and some of them advanced at the same time. New crevasses appeared, a massive avalanche buried the snout of Emmons Glacier on the north flank, and widespread floods washed out survey markers.

Valley glaciers like Nisqually and Emmons are closely watched because they are the first indicators of world climate changes. At temperate latitudes, continuous melting and re-freezing converts snow into glacial ice much faster than snow is converted in the cold wastes of Greenland and the antarctic. Thickening of the ice occurs readily in the narrow flows, and bulges pile up into sheer cliffs as much as 300 feet high—sometimes resulting in a sudden surge forward when the tension becomes too great. This has already happened in the far Northwest, and Nisqually now has an upstream bulge which has been in the making for the last fifteen years.

The records show that Temperate Zone glaciers have been growing while the climate below 70 degrees N. Latitude warmed. The probable explanation is that warming of the ocean surface causes increased evaporation, which, in turn, causes more snow to fall on the high mountains. Consequently, a glacier gains more at the top than it loses down-valley at its terminus through evaporation and melting. Eventually the glaciers in the far North will also grow, and the increase in cold, snow and glacial ice there will begin to change the trend in the South. As the air cools over the ocean, less snow falls in the mountains and the Temperate Zone glaciers shrink. The Rainier glaciers have grown as the result of a warmer climate and greater evaporation throughout northern Europe and North America. But the phenomenon may be in the process of reversing itself. For the last twenty years the mean annual air temperature for the world as a whole has dropped

.2 degrees F.

Warming and cooling—and glacial advance and retreat—may continue in cycles for thousands of years without causing a major advance of polar ice. Some scientists believe that another Ice Age would be possible only if the world warmed enough to melt the polar ice caps and enlarge the ocean surface of the earth. The sea would rise then and drown our coastal plains. Theoreticians calculate that such a development could take place some 10,000 to 50,000 years from now.

At depths of 20 to 60 miles beneath the earth's surface, the heat is apparently so great that rocks remain solid only because of the pressure of overlying strata. Wherever surface cracking relieves this pressure, a magma chamber of melted rock forms. The steaming hot liquid rises, sometimes spilling out through fissures over vast lava fields. From pipelike conduits come the violent eruptions of lava, ashes and steam that we call volcanoes, which leave behind huge conical piles. Sometimes lava cools in the vent of the volcano, forming a "plug dome" that stands long after the crater has worn away.

Every feature of the southern Cascades reflects volcanic fires, old and new. Mount Shasta's many past convulsions are recorded in the crags of its summit and in the sections exposed by glaciers that have moved down its slopes. Northward, in Oregon, rise Mount Pitt and the Three Sisters. Southward, the plug dome of Mount Lassen, more than 10,000 feet high, overlooks miles of cinder cones, volcanoes, and bubbling sulfur springs. This entire area was once a single volcano 1,000 feet or more higher than Lassen is today. But the top collapsed, creating a great bowl, or caldera, similar to the one occupied by Crater Lake in Oregon and to the Jemez Crater of New Mexico.

At the turn of the century, John Muir, who climbed to Shasta's peak, wondered whether the sleep of the Cascade volcanoes might not be broken one day, and lava pour down on the farms that seemed then so securely settled on the mountain flanks. "It is known," he wrote, "that more than a thousand years of calm have intervened between violent eruptions . . . volcanoes work and rest, and we have no sure means of knowing whether they are dead when still, or only sleeping."

Cinder Cone

In Lassen Volcanic National Park, lava fields from the last eruption of Cinder Cone in 1851 are colonized by few plants. A yellow pine that took root here lies in the foreground, toppled by wind in 1962.

Bumpass Hell

Sulfurous gases and steam rise from hot springs in Bumpass Hell, southwest of Lassen Peak. The basin is part of an old volcano which collapsed, forming a caldera, or large crater, 12 miles in diameter.

275

The Peppermint Peaks
*Tilted rocks that form the heights of Glacier
National Park are evidence of the stress and strain
within the earth for millions of years while the
Rockies were rising from the bottom of an inland sea.
The silt, sand, clay and limy mud of that ancient
sea have been changed by the heat of mountain-building
into varicolored rock strata. Above Sperry Glacier,
white quartzite alternates with red argellite in
a crumpled formation (below) nearly 3,000 feet thick.*

It was not many years later (in 1914) that
Mount Lassen began to erupt. Again, in 1915,
lava flowed and the vent became plugged. Hot
gases blew out this cork, and a destructive mud-
flow swept down the east flank. The smoke and
dust of the explosion were seen as far away as
Nevada. This part of Lassen Volcanic National
Park, one and a quarter miles wide and four miles
long, is still a "devastated area" in which trees
lie uprooted, their crowns all pointing away from
the peak. After 1915 the volcano was quiet, except for the formation of a summit crater in 1917
and one small eruption in 1925. Still, under certain weather conditions, trails of steam can be
seen emanating from fumaroles on the mountain
top.

The birth of Cinder Cone, which lies six miles
beyond a locked gate in Lassen Park, is remembered in Indian legend as a "fearful time of darkness, when the sky was black with ashes and
smoke that threatened every living thing with
death." Here the Fantastic Lava Bed filled the
center of a lake, dividing it in two, and a 700-foot
cone formed that sent up flares as late as 1851.
The loose cinders on the western side provide a
poor anchorage for tall yellow pines; they advance

like lonely sentinels from the forest edge, only to topple in the first heavy wind storm. Better equipped to survive is the lodgepole, the *Pinus contorta* of the high mountain moraines, which also takes hold here among the cinders of the lowlands, and bends but seldom breaks in the wind. Like the beds at Cinder Cone, hundreds of thousands of square miles of Western land are surfaced with cinders and lava which seem to have been deposited only yesterday. And these vast areas are practically bare of plant growth.

Not all of the molten magma in the earth finds an outlet. Some crystallizes below ground in the presence of trapped heat and volatile gases, to become granite. Denser than surface rocks, the bodies of granite form the cores of mountain chains that are pushed up and then ultimately destroyed innumerable times through geologic eras. Internal heat—the most likely cause of all kinds of mountain building—comes from the radioactive decay of uranium, thorium and potassium in rocks, and also from pressure. These are the forces that move the plastic mantle underlying the earth's crust, and fold or crack the more brittle

On Top of the World
Silhouetted against a glacial lake, mountain goats cross the tundra near Sperry Glacier. Wintering in lower valleys, where the kids are born in spring, they spend the summer climbing and feeding on the lichens, grasses and shrubs to be found at high altitude.

rocks of the crust.

It was once believed that the earth consisted for the most part of a molten core, surrounded only by a hardened crust that wrinkled up into mountain chains as the interior cooled and shrank. Now we know that the denser rocks of the mantle, capable of transmitting elastic waves from explosions and earthquakes, extend half way to the center of the earth, and overlie a relatively small core. This core is probably metallic and at least partly liquid. The mantle is completely solid except for small pockets of magma, the "hot spots" created by radioactive decay of fissionable materials. Mountains rise today, as in the past, under the influence of the magma chambers and waves, not because of any shrinkage that may

have taken place in the core. During block-faulting, the earth's crust may even expand. It ruptures and spreads apart as the mountain block tips and the valley block drops, forming a steep-sided depression.

After the cracking of great masses of crust into hundreds of individual blocks in the Great Basin during the Cenozoic Era, the shoreline of that period advanced from central Nevada to the Great Valley of California, which is today floored with sediments from the western, down-tilted surface of the Sierra Nevada. Nearly 400 miles long, this is probably the most extensive and magnificently sculptured granite block to be found anywhere on the planet.

Without mountain-building, the erosion of land

by wind and water would shrink continents to islands. During the age of dinosaurs, much of North America was inundated. It was from the bottom of an ocean trough that the immense Rocky Mountain system arose, stretching from Alaska to Central America and having a breadth of some 500 miles in Colorado, Wyoming, Idaho and Montana. Most of the energy of mountain-building during the Laramide Revolution of 135 million years ago was expended in wrinkling the bottom of the trough into great arches. But there were two later risings, one about 25 million and one about 12 million years ago, during which many of these arches cracked into pieces like the shingles on a roof.

The diverse character of the several chains in the Rocky Mountain system depends in large part on the materials from which they are made. In only two places did volcanoes play a major role in their formation. Lava flows built up the San Juan Mountain mass in southwestern Colorado and overwhelmed 18 successive forests on the Yellowstone Plateau in Wyoming. The petrified stumps are still standing, exposed, on the side of Amethyst Cliff.

Elsewhere, in the Central and Southern Rockies, the ranges were arched or tilted so high at various times that erosion has stripped away all but their granite cores. The lavas, shales and limestones that once covered them are now reduced to remnants on their outer flanks or within high valleys. In the Northern Rockies, the sedimentary layers in the old sea trough were comparatively thick. Here peaks are sculptured from candy-striped strata many thousands of feet in depth. At one stage of mountain-building, this rock crumpled and cracked, thrusting the bottom layers about 35 miles eastward over the top layers. Much of the Lewis Overthrust, as this formation is named, has eroded away, but some remains. The rock of Chief Mountain in Glacier National Park is at least half a billion years old, and you can clearly see where the mountain rests on the crumpled shales of a Cretaceous sea only 135 million years old.

There is a feeling of loneliness on a mountain top. In the great amphitheaters quarried out by the glaciers, few bird calls ring out. Practically nothing can flush the mute ptarmigan hen from her nest. Her summer plumage is an exact imitation of surrounding rocks, and if you should stumble upon her, as upon a stone, she will try to creep away rather than fly—depending to the last on coloration for protection on these barren grounds.

The echo of a falling rock is startling but also welcome, because it may mean that a band of mountain goats will appear suddenly and parade across that otherwise desolate stage. The mountain goat, like its European relative, the chamois, spends all summer ledge-hopping and playing in the snowbanks, sure of its footing on non-skid hooves. It is not a true goat, but a goat-antelope, native to the Rockies and the Cascades and recently introduced into the Olympic Mountains of Washington.

The high summer playground of the mountain goat is colder, windier, wetter and more severely radiated than the forested slopes below. Direct solar radiation is intense on the peaks because the thin atmosphere does not intercept and scatter the sun's rays as effectively as the turbid air at lower elevations. The effect is one of climatic extremes. The surface temperature of a stone, for example, may be more than 30 degrees higher during the day than at night. Consequently, the layered rocks crack and scatter, and on some slopes plants cannot take hold at all.

Each spring the snow melt in the Rockies follows the same patterns as in the Alps and the Himalayas. At Independence Pass, near Aspen, Colorado, the grass of the alpine meadow extends a series of green fingers onto one snowy ridge, while another ridge just beyond stands windswept and bare above pockets of snow. Austrian meteorologists in the Southern Alps have determined exactly how the melt-off progresses, given the degree of the slope and its orientation to the sun: first the steep south, then the gentle south and north, and finally the steep north slopes. Generally, the warmest slopes in the Northern Hemisphere face southwest; these heat up in the morning and dry out in the afternoon. They are populated only by plants with leaves which can stand extreme heating (a recorded maximum, in one instance, of 113 degrees F.) and which can roll up to conserve moisture.

Next to this mountain-top desert, there may be a glacier that imposes arctic conditions. On its

Tree Line
Whitebark pines mark "tree line" at about 10,000-foot altitude in the Grand Tetons of Wyoming. The trunks are twisted and their branches, growing on one side only, show the direction of the prevailing winds.

surface, life is limited to primitive, wingless insects such as the bristletail and certain worms, and to colonies of a red bacterium containing bromine, which spread in a faint blush over the snow. Below the glacier, a cold, ground-level wind flattens the grass, damages mosses, and stunts the growth of trees.

Where the snow lies heaviest and the glaciers are biggest, in the Pacific Northwest, the top of a mountain is almost uninhabitable at any time of year. But there are some strange exceptions. The summit crater of Mount Rainier, warmed by steam vents, shelters mosses and white-footed mice, and has been visited at various times by porcupines, ravens, rosy finches, pipits and eagles.

Like the tundra, the alpine meadow is a pioneer community. The growing season here is longer than in the arctic, but the days are shorter. Plants are mostly perennials, and it may take five years for a tiny cushion to reach one fourth of an inch in diameter. Nature gardens may be several hundred years old. Flowers such as golden bear grass bloom only once in five to seven years.

Similarly affected are evergreen trees at their highest altitude, where they are rarely able to produce cones and seeds. In the extreme cold of winter the tops are killed, and trunks bend under the weight of snow. The hot sun of early spring dries the needles more rapidly than moisture can be drawn up by the roots from the still-frozen soil. So the trees die back. But each species grows as high on the slope as it can, and every mountain region has its tree pioneers, standing gaunt and twisted at timberline. The whitebark pine grows at highest altitude. This conifer is to be found on all Western mountain ranges except the Olympics.

Two of the eight species of mountain heath native to the colder parts of the Northern Hemisphere grow in Rocky Mountain meadows—their needlelike leaves and tiny blossoms almost identical to the heathers of the Old World. These meadows also contain purple saxifrage, indistinguishable from the same plant in the arctic and in the Swiss Alps. There is, however, only a minimum of the thick arctic moss and lichen mat, underlain by permafrost, that extends as far south as Mount Washington, New Hampshire, in the Eastern mountains of the United States.

The rugged spurs of the high Western mountains are usually better drained, grassier, and have more flowers than those of the East. With increasing distance from Alaska, as the arctic flowers thin out, their places are taken by dwarf varieties of other species common to lower elevations. Thus, on the considerable tundra of the Colorado Rockies, the shooting star, the forget-me-not, and the red monkey flower mingle with such Colorado natives as the hawkweed, alpine aster, alpine spring beauty and alpine sunflower. This sunflower, the showiest of the floral display

above 9,000 feet, seems literally to spring from bare limestone rock. Its two- to three-inch golden head always faces east, making it a reliable compass at any time of the day.

A relative of purple saxifrage, the James Boykinia, is one of the plants that have adapted specially to rock slides, and takes root in chinks wherever there may be a crack large enough. It works its way in, forming its own soil with crumbs of rock and mulch from dead leaves, establishing its own niche in a vertical world. A prime advantage in such a clinging existence is inaccessibility to elk and deer, which are said to have a special liking for this plant whenever and wherever it can be found in an open meadow.

The mountain goat, the wild sheep, the deer and the elk all follow the gradual blossoming of plant life in the mountains from the lower valleys to the high ridges in spring and summer, and then disappear downward into the spruce and fir when the snow flies again. The pastures most heavily trampled and grazed during these up-and-down travels are the openings just below timberline, where trees serve to break the wind and hold moisture, and where each sunrise burns off a mantle of fog. Since hunters with rifles have reduced the herds—over the past century—sheep

Deer Park
As afternoon shadows lengthen on Hurricane Ridge, in Olympic National Park, Washington, a young mule deer pauses to stare uncertainly at a human visitor to this immense alpine meadow surrounded by firs.

The Pika

Sometimes called a whistling hare, or piping hare, a pika will reveal its presence in the mountains most often by its voice. A tiny animal, it weighs only four to six ounces. The pika has fur that closely matches in coloration the rocks which form its habitat.

and goats are seldom to be seen at the same time.

The mountain sheep traditionally range south into the Sierras, where the flocks have been decimated by epidemics of scabies contracted from domestic sheep. Only 400 were believed to be left on their ranges east of the Sierra crest by 1950. At the same time, the flocks of desert bighorns in the Mojave and Sonoran Deserts, protected on game ranges, were increasing in size. (One of the causes of the virtual disappearance of the mountain sheep may have been the growth of deer-food plants on cut-over slopes, bringing deer higher on the mountains than they had ever come before. The two are not compatible; when deer approach, the wild sheep will often withdraw to another feeding area.)

A rock slide, or talus slope, is inhabited by the tiny pika for much the same reason that the James Boykinia clings to its rock niche—for protection. In this case, the animal is trying to avoid being dug out by one of its many enemies—among them the marten and the weasel. The herbaceous plants and grasses that grow nearby provide the hay that this mouse-eared member of the rabbit family collects and spreads on its "doorstep" to dry, later to be stored within the rocks for the winter months. A pika's strange alarm call, disturbingly like a baby's cry, serves a second purpose in warding off other pikas from its home territory. The call may be heard from above timberline to the valleys thousands of feet below, wherever there are rocks in which the animal can make a home.

The marmots of the western United States are big, fat ground squirrels that frequently dig their burrows under single boulders or trees, right out

Les Trois Tetons

Legend has it that the Tetons were given their name by early-day French trappers and fur traders who saw a resemblance to the breasts of women in the three principal peaks of this spectacular mountain range. Here are the Grand, Middle and South Tetons as seen reflected in a clear glacial pool from an unusual observation point on the western slope.

The Marmot

Surrounded by glacial ice, a marmot warms itself basking in the sun. This lumbering rodent, which weighs up to twenty pounds, dens in the rocks but needs some soil in which to burrow. There the animal hibernates during the winter, not storing a supply of food as does the busy pika.

A Glacial Canyon
Fir forests reach upward through talus slopes on the sides of Cascade Canyon in the Grand Tetons of Wyoming. The U-shaped canyon was gouged out by Ice Age glaciers that have since disappeared.

in the open. But they, too, have discovered the virtues of rock slides that their enemies, the badgers and wolverines, find difficult to dig into. In their holes the marmots, fattened on vegetables and insects, become dormant for much of the year. (The period depends on geographical location and elevation; an animal living in a lower valley in California will emerge from hibernation a month or two earlier than one living in the Sierras.) Sometimes the marmot and the pika inhabit the same mountain slope, as in Glacier National Park in Montana. For the human visitor, even without a visual clue, the clear, loud whistle of the marmot identifies it unmistakably. Then the great, lumbering animal may appear on a rock ledge somewhere above you, curious but ready to flee at the first sign of your approach.

Both the marmot and the pika intrude into the mountain forest, which is a more competitive zone populated by a great variety of life, much of it occupying highly specialized niches. Along hiking trails, the golden-mantled ground squirrel becomes familiar as a cute beggar during the tourist season. But its real place in the pine and fir woods has nothing to do with peanuts and bread crumbs. This squirrel can co-exist in a patch of forest with a chipmunk that eats exactly the same foods because the two animals have different habits. The squirrel gathers nuts and seeds on the ground; the chipmunk climbs the trees to harvest them before they fall. The squirrel puts on thick layers of fat to provide itself with energy during hibernation; the chipmunk stores food and does not hibernate. Neither interferes with the other in nature. When man enters the high-mountain environment, he often destroys such fragile relationships.

Man-made erosion in the mountains is some-

A River Canyon

The Black Canyon of the Gunnison River in Colorado provides a dramatic example of a river cut. In some places it measures only 1,300 feet from rim to rim, and narrows to 40 feet at the bottom.

what minimized by the prohibition of forest-cutting on steep slopes, which protects watersheds from washouts. The noise and pollution of automobile traffic is held down by the restriction of thousands of miles of back-country trails to pack and foot travel only. The higher you go, the more restrictions you find. A mountain itself sets up barriers; beyond a certain point, it is just not economically feasible to replace bridges or to clear fallen trees that may block trails for years after a snow avalanche.

To reach the Snowmass Wilderness Area, near Aspen, Colorado, you must leave your vehicle and cross fallen aspen and cold, raging mountain streams for miles on foot before the old jeep road ends at a sign reading "No Horses or Vehicles Beyond This Point." And yet, unprecedented inroads are being made into this wilderness. At the boundary of the National Forest, eight miles

northwest of Aspen, a resort community is rising that, in the next 15 years, will house some 17,000 skiers. Of the 10,000 acres slated for development, 6,500 acres are owned by the Forest Service, which is leasing the land to two large investors in Los Angeles for a percentage of the ski-lift profits. Entire new industries are growing up around the invention of a small, inexpensive, gasoline-powered vehicle, the snowmobile, that will make winter resort living and travel in the high mountains much easier than ever before.

Gradually all-season, two-lane, paved roads are replacing partly-paved, one-lane roads in mountain areas, and the concentration of people that the new roads will attract to certain places close to large cities is causing concern. Here the Department of Agriculture's Forest Service comes into direct conflict with the Interior Department's National Park Service, whose administrators are

285

mostly dedicated to preserving land from such intensive development. Secretary of Agriculture Orville L. Freeman and Secretary of the Interior Stewart L. Udall often opposed each other, sharply and publicly, on such issues—recently on the question of building a new road to Mineral King Valley, in the high Sierras, where the late Walt Disney's organization plans a $35-million complex of lodges, parking space, ski lifts and other facilities under the auspices of the Forest Service. Six miles of the proposed road would run through Sequoia National Park, which is downstream from Mineral King on the Kaweah River. And Secretary Udall, no doubt thinking of the outdoor slums in Yosemite Valley, flatly stated that he could not see how it would be "humanly possible to crowd 20,000 people and parking lots containing 8,600 autos into this narrow valley without serious damage to the quality of the water in the river and serious air pollution as well."

The pioneer de luxe paved road in the high mountains, Trail Ridge, has been taking sightseers on a breath-taking tour of the northern Colorado Rockies for a little more than a quarter of a century. Unlike the Arapahoes, the Utes and the "mountain men" who once passed this way slowly and painfully on foot and on horseback, the modern traveler can see many of the 84 peaks in Rocky Mountain National Park which are over 11,000 feet high, and can cross 11 miles of alpine meadow, all in half a day. He doesn't even have to leave his car. If he does, he will probably step on the flowers, and thus help destroy the slow-growing alpine meadows on either side of the road —a wantonness greatly deplored by the Park Service—and he will do even more damage by feeding the chipmunks and ground squirrels. These will multiply and eat more alpine flowers than the tourists will ever step on. At Sunrise Meadows on Mount Rainier, hordes of these little animals descend on gardens planted right in front of the exhibition center and efficiently cut down the blossoms in between taking handouts from the visitors and biting fingers.

Wherever traffic flow is heavy, the ecology of an area tends to become unbalanced; wherever it isn't—as on 300 miles of back-country trails in Rocky Mountain National Park and on 1,000 miles of trails in Glacier National Park—there is less change. And the largely inaccessible heights —the North Cascades, for example—are left virtu-

The Beaver Pond
*Lines of aspen logs, such as the large
tree neatly severed from its stump in the
foreground, barricade the water within this
beaver pond in the Centennial Range. The
beaver stores some branches underwater for
consumption during winter months of ice.*

A Working Beaver
*In a beaver pond at Red Rocks, Montana
an industrious animal moves a stick into
position to reinforce the dam that holds back
the waters of a shallow creek. It uses
both mouth and paws to pick up materials and
tamp mud into place to repair any leaks.*

ally unchanged by visits from a few hardy camp-
ers and climbers.

The North Cascades National Park, created on
October 2, 1968, is made up of approximately
680,000 acres in northern Washington that formerly belonged to the U.S. Forest Service. It lies
within a great mountain fastness that is still
owned and managed by the Department of Agriculture, either for limited use as wilderness, for
recreation, or for commercial timber, grazing and
mining. Of the entire 6.2 million acres recently
surveyed, very little has been developed; 28 per
cent is lonely high country—grassland, alpine
meadows, bare rock and fields of snow and ice.
Two-thirds is uncut forest, classified non-commer-

cial because of steep slopes, fragile soils, or inaccessibility. But when bulldozers gouge out new
mining roads—for Kennecott Copper Company
and others that will follow—and when a paved
highway is laid into the remote heartland of the
Cascades, a human tide will surely rush in.

Rocky Mountain National Park is the site of a
gigantic mountain engineering feat. The Colo-
rado-Big Thompson project at its western limit
diverts water from man-made reservoirs and from
Grand Lake, all in the Pacific drainage, and routes
it through tunnels to the east side near Estes Park,
where it irrigates farmlands on the plains. Centuries before men's engineering, however, the

288

A Beneficiary
A spotted frog floats beneath the surface of the deep, quiet pond built by a beaver and its mate. Often inhabiting the bogs on the valley floor, this frog may also be found in beaver ponds at altitudes up to 10,000 feet in the Montana mountains.

beaver began to alter the character of mountain streams. Descending from Trail Ridge on the eastern slope, a stair-step pattern of beaver dams captures cold, swirling waters and turns them into warm, still pools. Many of the lodges are uninhabited now. But the fact remains that the largest rodent in North America has survived the trappers that sought its soft, valuable pelt in the Colorado, Montana and Wyoming Rockies and throughout the Pacific Northwest.

In ten years, one pair of beavers may produce as many as thirty offspring, and these will spread in perhaps half a dozen colonies up and downstream. In their ponds they impound the spring runoff, and hundreds of acres of meadow are sub-irrigated and kept green through the summer —benefiting vole and hare, coyote and red fox, hawk and owl. In each pond grow the water plants on which the muskrat lives, as well as a variety of insect food for frogs and fishes.

There are not likely to be many trout in a beaver pond, because trout prefer colder, more turbulent waters. For this reason the dam-builders are unpopular with fishermen. To a beaver, however, the dam probably means the difference between survival and death by starvation in the winter. In the mud bottom of the deep pond that forms behind a dam, he can cache away enough green twigs and sticks to tide him over the long months during which lodge and stream are ice-bound. All around the pond, he cuts down aspen and willow—as a good forester would, for "sustained use," leaving one patch to continue its growth while another is downed. Eventually, after many years, he will move up or downstream in search of new stands.

Constant building and repairing of his lodge seems to satisfy an inner need in the beaver. But some beavers will do no more than dig a den with an underwater entrance in the river bank. In valleys where beavers have been hunted almost to extinction, only these den builders, whose habits are inconspicuous, have survived. European beavers living along the wooded banks of the Rhone and in other river basins heavily populated by people have not built any lodges since the Middle Ages, though beavers in the remote taiga of Russia and Scandinavia still construct them.

Along U.S. mountain streams, after a colony of beavers has moved—or disappeared because of trapping—new aspens quickly grow up. The older trees send out runners and populate the cut-over areas with shoots. Thus the roots of nearly all the trees in a grove may be connected, and the entire grove will sometimes fall down as one, uprooted by wind or rock slides. In the wake of fires, aspens colonize valley slopes in the same manner, before the seedlings of the spruce and fir can take hold.

The fir usually colors the north slopes of a mountain valley dark green, as on a vegetation map, in contrast to the light green of the quaking aspen on the south slopes. But each slope has its variations; the steeper parts of the north side are often snow-slide areas, striped with lines of small aspens only two or three feet high and bent downhill. Streams running down the mountainside serve to irrigate small level areas below—grassy meadows, groves of willow as tall as a man, and stands of larger aspens. These flatland aspen groves contain more ferns and flowers, birds and other wildlife than those on the mountainside.

Nowhere are the mountain-and-marsh habitats more extensive than in the Jackson Hole, Yellowstone, and Red Rock Lakes region of southwestern Montana, northeastern Idaho, and northwestern Wyoming. In the early part of the 19th Century, this was the southern limit of the breeding range of the then-abundant trumpeter swan.

For nearly 100 years, the swan-skin trade of the Hudson's Bay Company reduced the far-flung colonies on mountain and prairie lakes. Today Red Rock Lakes, just west of the Yellowstone in Montana, harbor the biggest colony of trumpeter swans south of the Canadian border. Since 1935, the shallow ponds and sedge-filled marshes of the area, fed by the high escarpment of the Centennial Range, have been set aside for the preservation of the swans.

This federal refuge, remote from the mainstream of 20th Century life, preserves on some 40,000 acres not only trumpeter swans but also many other representatives of the original wildlife of the Northern Rockies. The nearest town is Monida (population 20). The only roads to the refuge are little more than truck or jeep tracks. When snow locks in the 6,600-foot-high Centennial Valley early in the fall, even these trails become impassable except by snowsled, which brings in supplies for the swan-keepers and grain to feed the swans. Since the refuge was established, the birds have almost ceased migrating. All but a few gather at two warm springs that remain ice-free during the winter. The rest fly a short distance to the traditional wintering ground of Ontario trumpeters, the North Fork of the Snake River in Idaho. These waters, too, are kept open by warm springs, even though the air temperature may drop to minus 30 degrees F.

When summer comes, a handful of people wander through the refuge on their way to fishing camps, or perhaps just to look at birds. Between marsh and mountain there is a small campground among the aspen groves on the shore of Upper Red Rock Lake. To the north, under the backdrop of the Gravelly Range, flotillas of ducks, coots, and geese cruise about over some 10,000 acres of watery wilderness. For a glimpse of the nesting swans you will need a telescope, but the sandhill cranes—which are abundant here—are not so shy; they announce the dawn of each day with a cacophony of loud bugling immediately overhead.

To the south, dark forests of Douglas fir and lodgepole pine sweep abruptly upward to meet alpine wind timber, and above are the snowy, barren peaks of the Centennial Range. Some of the wildlife species indigenous to these high and distant zones are only memories. Mountain sheep have been gone for more than half a century.

The Lynx and the Hare
Surprised during a meal, a Canada lynx crouches over the remains of its prey, a snowshoe hare in brown summer fur. Lynx populations have been found to fluctuate cyclically, following cycles in the populations of hares.

Marten, fisher and grizzly bear have not been seen for decades. The grayling, an arctic game fish once plentiful in the clear, cold streams, is on the way out. But the peregrine falcon, often observed diving from the cliffs, is well-fed on ducks and other waterfowl of the marshlands. Winged hunters of the forest—the goshawk and the sharpshinned hawk—swoop after squirrels and songbirds, and the shadowy lynx still haunts the trail of the snowshoe hare.

The mountainside maintains, in a narrow belt of conifers, some of the primeval rhythm of life— the rise and fall of populations of prey and predator—as this rhythm exists in the spruce and fir forests to the north. The numbers of the lynx fluctuate every nine to eleven years, following the cycle of its main prey, the hare. In Canada, the population density of the snowshoe hare varies over this period of time from as few as one or two per square mile to as many as 3,400 per square mile. Such a great change in food supply has a drastic effect on the predatory lynx, which becomes most numerous only to have hares suddenly "crash." Then thousands of lynx starve to death. The records of the Hudson's Bay Company from 1821 to 1934 amply document the cycles, one dependent on the other. A good catch of rabbits—on the average, 115,000—was always followed by a good catch of lynx—as many as 60,000—the next year. Since the turn of the century, many hunters and naturalists have observed the same phenomenon. Ernest Thompson Seton, traveling the Athabaska River Valley during the "bad rabbit year" of 1906–7, saw a dozen emaciated lynxes hunting in this area alone, and he found the carcasses of a dozen more in the woods.

Why does this happen? No one knows exactly,

although there are many theories involving sunspots and weather and plant growth—all of which are cyclic but not necessarily related to the cycles in animal numbers. A partial explanation can be found in the tendency of a population to increase until there is not enough space for the individual to function normally.

The hare prefers the aspen or the willow bog which provides the most food and cover. There it stays when the population is low. As time goes on, the fertility of the does increases and there is little mortality among the well-fed, healthy young. Litter size grows from two to three or four young, and a doe may have two litters in a season. Meanwhile, the hares spread out through the undergrowth of the forest. Even though they roam far, however, they cannot escape from other hares. The animals become more aggressive, and the struggle for dominance impairs the breeding condition of both males and females. In some does conception is blocked. In others, there is resorption of embryos. Many that do become mothers are unable to nurse their young, or for some other reason abandon them altogether. What began as a nervous condition under the stress of crowding—and in the case of hares, more than two per acre is a crowd—now becomes hypoglycemia. This shock disease (actually a deficiency in blood sugar) sets off a chain of physiological responses affecting weight, growth, resistance to disease, reproduction and water balance, among other things. For many hares, it ends in convulsions and death.

In recent years, studies of lemmings, voles, rats, muskrats and snowshoe hares, both in the laboratory and in the wild, confirm that these die-offs are due not to epidemic diseases, as previously thought, but to stress and shock. It is of some interest to pathologists concerned with the effect of crowding on people that hypoglycemia in mice can be arrested early in the aggression stage by injecting a tranquilizer. Our human society is a crowded one, and one that is increasingly dependent on drugs, from aspirin to LSD.

On Upper Red Rock Lake, the almost continuous forest of aspen is broken by shorelines of sedge and rushes, which give way to a low-lying bog dotted with treacherous potholes and crisscrossed by the trails of moose. Named "muswa" by the Chippewa Indians, this northern species is the same animal as the European elk. Its range roughly corresponds to the spruce and fir zone, but most of its food grows in the aspen forest and the bog. In summer, moose bed down in the tall sedge, or else browse, partly submerged, on waterweed, pondweed, and other aquatic plants. In winter, they paw through the snow to reach young evergreens and the bark of deciduous trees, and help themselves to the refuge haystacks. The annual count here varies from 20 to 35 moose, depending on the number of drifters from surrounding mountain ranges. In some years hunters are allowed to shoot one or two as a population control measure.

Like the deer and wapiti, moose do well in disturbed areas overgrown with shrubs and young saplings, where their populations often increase beyond the capacity of the land to feed them. Where this has happened, as on Isle Royale in Lake Superior, bogs have been trampled into mud holes. For the last several years, however, a population of 600 moose on the island has been held in check by some 30 wolves, "gray ghosts" that only recently made their way across the frozen lake from Canada. On this island there is currently a balance between the large carnivore and its prey that exists nowhere else in the United States, except in parts of Alaska. Whether the balance will continue depends on preservation of the moose habitat.

This habitat is imperiled, ironically, by fire control. Major forest fires have been prevented on Isle Royale since the turn of the century, and tall, old spruce are shading out the alder, birch and willow stands which provide browse for the moose. The prospect, clearly, is starvation soon to come.

All predators have, in winter, an incomparable ally in the deep snow. On the upper slopes of the

Wide-ranging Moose
On a hot day in August, bull moose slog comfortably through the wet sedge at Upper Red Rock Lake in Montana. The big bull in the lead wears a red necktie and ear tags put on him in March of the same year by a biologist on the other side of the Centennial Range, at a station in Idaho. The animal's odyssey is not unusual; older bulls customarily leave their mates in the spring, as soon as newborn calves are able to walk. Then for some months the bull moose is a wanderer.

293

Western mountains, snow drives the big game into the valleys, where large herds are besieged by coyotes, bobcats, mountain lions and wolves. The Indians knew about these congregations, and were willing to brave cold, avalanche and storm to follow the game at this time of the year, each tribe jealously guarding its hunting grounds. The journals and paintings of Father Nicholas Point (1840–1847) vividly record how the hunters drove the animals into snowbanks where they stayed, helpless, until slaughtered. After the hunt, wolves would boldly close in around the fires and set up a chorus of howls while the skinning was being done and the meat smoked. The Gros Ventres Indians were among those who made war on the wolves, capturing them by the dozen in pit traps for their warm fur and their fat, which they used for dressing hides. A little more than a hundred years later, other men who succeeded the Indians in the Rockies would hunt, trap, and poison wolves almost to the point of extinction.

In the high plateau country west of the Rockies, every stream, from the mighty Colorado to the smallest tributary, has lived twice—once as an old river, meandering across flat plains, and again as a youthful stream, gouging deep, V-shaped canyons from the uplifted plateaus of the present era. At least 2,000 feet of the uplifting occurred in the coldest period of the Pleistocene Epoch, when Wisconsin ice capped Grand Mesa, the Aquarius and other plateaus that today stand 11,000 feet or higher. They may still be rising, keeping the river cuts narrow and young-looking in some places. But elsewhere a combination of desertlike climate, wind and flash floods has carved out broad valleys full of monuments, arches and pinnacles.

In the mile-deep Grand Canyon are the most spectacular displays of these works of erosion, the harder rocks giving shape to the cliff tops and spires and forming the steps of terraces which descend some 2 billion years into the history of the earth—each colored band furnishing a re-

The North Rim
From Point Imperial (8,801 feet), the highest spot along the rims of the Grand Canyon, a maze of pyramids extends for miles across to the Coconino Plateau.

minder of changing climates and evolving animal life. Some of the oldest and some of the newest layers are missing. At the upper (latest) end of the time scale, marks of the age of dinosaurs and the age of mammals have been mostly eroded away. At the lower end, the river chasm contains no sign of the dawn of life, although rocks of similar age in Canada and the Sierras have yielded the fossils of primitive bacteria and algae. In these darkest depths, Precambrian mountain building long ago recrystallized the rocks into solid granite, gneiss and schist, like the peaks of the Tetons and the Front Range of Colorado.

Few places on earth are as isolated as the cool green forest of the Kaibab, on the North Rim of the Grand Canyon, 9,000 feet high and almost completely protected from surrounding sagebrush flats in Utah by impassable rimrock. This lofty plateau has been geographically remote from the South Rim since the canyon was cut thousands of years ago (and still is 214 miles away by road). Consequently, its ecology is unique.

The Kaibab receives about ten more inches of precipitation than the South Rim each year—mostly snow, which lies deep in the pine, fir and aspen forests near the rim from September to May. A thick layer of porous limestone underlies the gentle arch of the plateau. From its springs come an abundance of water; Robber's Roost Spring alone spills fifty gallons a minute into the aspen parkland, a summer feeding ground of the Kaibab deer herd. The narrow, grassy valleys of the region are swamped with water until the dry season, in late spring and summer—the reason being that only three inches under the clay-loam of the open fields is hardpan.

In the ponderosa pine cavorts a big, gray,

tassel-eared squirrel indigenous only to the Kaibab North. Many eons ago, as the canyon widened, a population of tassel-ears became isolated from others on the plateaus to the south. Gradually its appearance and habits changed, and a new species was formed. Today the Kaibab squirrel has a black belly and a pure white tail; Abert's squirrel, its relative on the South Rim, has a white belly and the upper side of its tail is black. And there are other, more subtle differences. The Kaibab squirrel dislikes contact with people, and avoids areas crowded with chickarees, golden-mantled ground squirrels and chipmunks. Or so it would seem. For more than a decade, it has practically deserted the National Park campground on the North Rim, and now builds its nests high in the pines around Jacob Lake. Perhaps it finds a better supply of cones and of young, tender, inner bark in the second-growth forest of the deserted lumber camp there than in the aging pine groves of the park, where no lumbering has been pemitted.

On the Kaibab North, the conditions for population imbalances among wildlife of all kinds started with the introduction of cattle and sheep before the turn of the century. The land was overgrazed when President Theodore Roosevelt created the National Game Preserve on the North Rim in 1906. Unfortunately, grazing continued, especially in the deer parks. And in the next decade federal hunters exterminated the timber wolf, virtually eliminated the mountain lion, and killed thousands of coyotes and many hundreds of bobcats.

Without predators, the deer population rose from 4,000 in 1906 to a maximum of 100,000 in 1924. In the vital wintering places on the west slope, the cliff rose and other shrubs were high-

Twisted Aspens
On a hillside above VT Park, the slim forms of aspen trees are bent in many directions, like a company of exercising ballet dancers. Heavy winter snows have molded the pliable trunks into these shapes.

A Rare Squirrel
Flashing its white tail, a Kaibab squirrel pauses on a pine branch in the Kaibab National Forest. This rare species, which has long ear tassels in winter, is found only on the Kaibab Plateau.

A Squirrel Predator
At Greenland Spring, in one of the grassy parks of the Kaibab Plateau, a goshawk lands briefly for a refreshing summer drink. This bird is a skillful hunter of small mammals—especially squirrels.

lined. Even sagebrush did not escape the teeth of cattle, sheep and deer. In the winter of 1924–1925, 75 per cent of the fawns died, while thousands of older deer starved to death. Fawns are always the first victims of starvation because they can't reach as high as larger deer for the twigs and buds of bushes and trees. Next to perish are the does and yearlings, and finally the big bucks.

Trapping and transplanting of fawns was tried here, in an effort to save them from sure death on the range, but few survived. It was here, too, that the great Arizona deer drive was attempted. A group of local businessmen sent Indians and cowboys to beat the deer out of the brush and herd them across the Grand Canyon to the less populous ranges of the South Rim. The men lost their way in the woods and the deer all slipped through the line. Then, in December of 1928, the federal government hired professional hunters to

reduce the herd; they came in and shot 1,124 deer. Local indignation at this action was so great that the Governor of Arizona and the game warden of the state sued the government in Washington, and won, in the United States District Court. In spite of a later reversal by the Supreme Court, systematic culling of the herd was stopped and has never been tried again.

During the next two years, a change in the weather—good summer rainfall and heavy winter snows—revived some of the vegetation and brought the deer herd up to 14,000. But the relief was temporary. In 1947, a joint study by the state and federal governments found the range to be as poor in condition as in the mid-twenties. Once more the parks were becoming dust bowls, the trees and shrubs highlined. And meanwhile another large deer population was building up.

Only then were cattle numbers drastically re-

A Roaring Lion
A mountain lion screams in a Western forest. Two of these animals can kill 50 deer in a year. In some places, where the lions have been allowed to survive, a balance is kept between predator and prey.

A Leaping Doe
A frightened mule deer soars over a fallen tree. Because most of the mountain lions have been killed off on the Kaibab Plateau, human hunters have been unable to hold down the deer populations.

duced and sheep taken off the range completely. At last, steps were being taken to repair the damage done to the deer. For the first time it was realized that the "disease" of overpopulation must be treated, not just its symptoms—malnutrition and starvation.

Mule-deer numbers had always fluctuated with the weather and the browse conditions. Their migration patterns in winter had been the same for centuries. It was their habit to congregate in sheltered regions on the North Rim of the canyon or drift to the sagebrush and cliff-rose areas on the west, ignoring other feeding places that were deep in snow. The worst winters restricted these movements and caused crowding and starvation. But in two good years the herd could almost double in size, partly because the big predators were gone. In twelve years, from 1906 to 1918, some 600 mountain lions had been exterminated,

and the remnant population of about 70 (or one lion to every 200 deer) was not enough to check the high survival rate of the fawns and yearlings. The herd was increasing steadily in spite of losing one or two individuals every week to a lion.

Game management through hunting was not the answer. In the years of the population rise, hunting permits doubled, tripled, and then quadrupled. By 1950, it became necessary to open the deer parks to hunting in the summer, in addition to the November "trophy" hunts. Permits for late season shooting now included does. From 1951 to 1954, the number of deer permits was increased from 5,000 to 12,000, with no restrictions. But it was too late to avoid the crash. That winter, according to an Arizona Fish and Game survey, 18,000 deer starved. The plateau that had provided Paiute Indians and Mormon settlers with abundant game was again poor in deer. The mod-

ern hunter found himself in the position of the Northern lynx, a predator following a cycle—and game managers were hard pressed to find a way to stop the cycle's turn.

In the last decade, an attempt has been made to hold the line at about 10,000 deer. Weather sometimes helps and sometimes hinders. A dry year may lower fawn production. But in a wet year, when deer numbers are rising and control is crucial, many hunters either fail to show up or leave without their quota, because of stormy fall weather. Paradoxically, the mountain lion is still tracked by agents of the Division of Wildlife Services, even though every track is so rare as to be a matter of record. On hundreds of acres of forest felled to create more browse for the deer, coyotes are controlled; and as a consequence, rodents eat the new growth and plantings. Though reduced in numbers, livestock are still concentrated in places where they interfere with the game, as in VT Park (the name comes from an old cattle company brand), where it is not unusual to count 100 to 250 head of cattle within a few miles. This is the main summer range of the deer, a narrow green corridor in the aspen forest through which every tourist passes to reach the North Rim of Grand Canyon National Park. If you stop to photograph the famous herd there, you, too, will notice how crowding leaves its mark on the open forest. The Kaibab is still a scarred habitat, though it is in the process of attaining a new equilibrium.

So varied is the topography and vegetation of the Grand Canyon that no one can see it all from a single vantage point. On the high, rocky peninsulas of the North Rim, vistas of the painted desert fade away to the east and volcanic peaks loom to the south—impressive panoramas, but giving no clue as to what lies below. A descent into the canyon brings a realization of its tremendous depth and also of the disparity among its several climates. From an Upper Sonoran Zone of piñon pine and juniper, the South Rim dips half a mile to the great plateau called the Tonto Platform, through which the river has cut its V-shaped chasm, 1,500 feet deep.

This second rim has a Lower Sonoran, or hot-desert climate, with a flora and fauna all its own —burrowbush, cactus plants, yucca, and mormon tea, the last-named a strange, leafless holdover from Paleozoic deserts, whose jointed branches, arranged like candelabra, do the photosynthetic work of leaves. Here are birds not seen above— the desert sparrow, the lazuli bunting and the long-tailed chat. At rare intervals, cottonwoods cluster around a spring. Until recently, all the water used by visitors to the arid South Rim was piped up from one spring on Bright Angel Trail. But the current pressure of about a million visitors a year has made it necessary for park authorities to lay another pipeline, by helicopter, *across* the Canyon, to bring water from the North Rim.

Only from the air can you see the bold outlines of the canyon's structure, the origin of which has been a subject of controversy for some 72 years. Here the river did not simply cut downward as the plateaus rose, an explanation offered by the canyon's first explorer, John Wesley Powell, in 1896. Geologists have found that the folding and some of the faults that raised the canyon area occurred long before the Colorado River existed. A major mystery has been the discrepancy in age between the Upper and the Lower Colorado. At its source in the Colorado Rockies, the river is at least 65 million years old; but measurements of the potassium-argon ratio in volcanic sediments of the river below Lake Mead, where it follows the border of Arizona, recently showed an age of only 10 million years. The latest hypothesis is that two river systems existed during the Cenozoic Era, one to the east and another to the west of the Kaibab, which must have been one big plateau occupying the entire canyon area. From the east side, the Ancestral Upper Colorado turned toward New Mexico and drained into a lake. The Hualapai drainage system, which originated on the dome of the Kaibab, flowed west. As a waterfall will consume the ledge over which it tumbles, the headwaters of the Hualapai apparently worked their way backwards to capture the Ancestral Upper Colorado. After that, the force of the new Colorado—two rivers in one—cut deeply through the rocks, slicing apart the Kaibab and forming the Grand Canyon.

The River Slowed
Sand bars line the edges of the Colorado River as it winds through the eastern end of the Grand Canyon. Once a raging torrent, the river flow has been reduced by dams.

The South Rim
O'Neill Butte towers over the Tonto Platform, a desert region about 3,000 feet below the South Rim of the Grand Canyon. The butte itself is a two-hour hike from the rim, on the start of a trip into the abyss that takes a full day.

Flood Waters

Subtropical vegetation surrounds Havasu Falls, in a side canyon at the western end of the Grand Canyon. Ordinarily the clear, blue-green waters making the drop create a delicate spray that fills the travertine pools below. But within an hour after a cloudburst muddy waters (below), augmented by overflow from the South Rim, rage through this lower canyon. Such floods have left great sheets of "flow rock" overhanging the falls.

For millennia, the rock debris from landslide and flood has supplied the rushing waters of the Colorado with cutting tools—sand for scouring and boulders for pounding, to continually widen and deepen the canyon. Once an average of half a million tons of mud and sand, and probably an equal load of rocks, were carried past any given point in the river's winding course every 24 hours. But in 1963, Glen Canyon Dam, upstream from the Grand Canyon, stilled the wild river considerably in the span between its gates and Hoover Dam, a landmark on the Lower Colorado since 1936. Today, every ounce of water not absolutely necessary to meet power and irrigation commitments from Hoover Dam down to the Gulf of California is trapped behind Glen Canyon Dam, so that its generators can produce power. The river can be turned off and on at will. Before the Upper Colorado was tamed, the water ran at 100,000 to 200,000 cubic feet per second; now the maximum is about 10,000 to 20,000 cubic feet per second. And whenever Lake Powell must be filled, the flow is cut for a time to a relative trickle—1,000 cubic feet per second.

What has this done to the Grand Canyon? Nothing that can be seen from the rims. But far below, there have been important consequences. Above the older dam on the Lower Colorado, the rapids were silted over for a distance of 80 miles when Lake Mead was high, a decade ago. From Separation Canyon south, the "river" has become essentially "the lake," although rapids emerge whenever the lake level drops. Below the new dam on the Upper Colorado, the flow is no longer sufficient to flush away the rocks and gravel brought down from the high country by tributary streams. Just before the river enters Grand Canyon on the east, runoff waters from the palisades once cut gorges as deep as the canyon's own. Every junction became a rapid, piled with boulders. Because the river was squeezed into a narrow channel, it might rise a hundred feet after a heavy spring runoff or a summer shower. At low water jagged rocks would protrude, while at high water white caps boiled over a brown flood, heavy with silt.

By preventing such floods, the dam will cause gravel and boulders to pile up at the mouths of the innumerable side canyons, at Marble Gorge and within the National Park, eventually impounding the river behind a series of boulder dams which will form impassable rapids. No matter how much of the river silt is held in Lake Powell, the run-off from tributaries below Glen Canyon Dam will still find its way into the park, laden with the products of erosion.

Glen Canyon Dam opened the way for countless new plans to encroach on this one remaining unused part of the Colorado River. Halfway through the Canyon, at the entrance of Kanab Creek, engineers from Los Angeles have envisioned a long tunnel that would divert 91 per cent of the water to a power station and reservoir 45 miles away. It would not, they say, damage the scenery. On the eastern boundary of the park, the proposed Marble Gorge Dam would back water almost to the foot of Glen Canyon, destroying an area that many people feel should be included in the National Park.

The "marble" of Marble Gorge actually is a sheer cliff of Redwall limestone, one of the most distinctive formations in the Grand Canyon. Its color comes from iron oxide washed down from the red Supai and Hermit formations above. And it is full of solution holes—large and small caves from which streams fall, as from garden steps during a heavy rain. Beside and below them, ferns spread in profusion, and wildflowers grow in the hanging gardens. No one knows how much water might be lost through these holes, if and when Redwall Canyon, Vesey's Paradise, and the sand beaches are inundated and all their animal inhabitants drowned. Probably what happens here would be essentially what has happened in Glen Canyon.

Before the dam was closed there, the University of Utah made a study of the ecology of the canyon depths and found that at least 96 birds and 41 mammals would be affected by the flooding of willows and cottonwood in river bottoms and side canyons, and by the inundation of hanging gardens, springs and seeps on steep cliff walls. The introduction to the University's publication on mammals gives some clues to the variety of ways in which the life of the canyon floor is changed: "When the reservoir becomes filled, the water for the most part will stand against barren cliffs, slickrock or barren desert hillsides. All habitats that currently exist in the bottom and those up to several hundred feet will be inundated. . . .

Solution Holes

On the face of a cliff, hundreds of miniature caves are formed in soft sandstone by the solution of the cements that once bound grains of sand together. The solvent is water, seeping through cracks from above.

Beavers are doomed. Their present saturated populations will disappear because their entire area of food and den sites will be destroyed. Some may move up the tributaries, but here the food supply is scarce. Others may move upstream, if possible into Cataract Canyon, and on into the Green and Colorado Rivers. These areas, however, are already heavily populated. Deer will likewise disappear from the area, largely because of the dearth of food. . . . The rodents that presently inhabit the terraces, talus and hillsides will disappear because of the lack of food and homesites. . . . Some few inhabitants of the cliffs and ledges and those who are able to live in the adjacent shallow soils may persist."

At Marble Gorge, much of the bighorn sheep range is above the expected high water mark. Not so, however, in the area marked for Bridge Canyon Dam, just below the National Monument on the west. This new dam would be the highest in the Western Hemisphere, and the reservoir behind it would flood the entire length of the Monument and thirteen miles of the National Park.

The tragedy of the situation is that the proposed dams would not even supply water. They are power projects that the Bureau of Reclamation believes it must build to sell electricity and thereby help finance the Pacific Southwest Water Plan. This plan is certain to be approved by Congress, in one form or another, within a year or two. So many doubts have sprung up about the earning capacity of Marble Gorge Dam, however, that it has been temporarily shelved. And Bridge Canyon, though still on the agenda, is being re-evaluated. It has become evident that the amount of water the Colorado River can provide has been vastly overestimated ever since the early 1920's. The Bureau of Reclamation has never been able to fill Lakes Powell and Mead full enough to operate the Glen Canyon and Hoover Power Plants simultaneously at rated capacity. Either Powell has been lowered to provide a minimum operating head at Hoover, or Hoover has been kept below rated capacity in an attempt to raise Powell's level. Meanwhile, electricity must be bought from other suppliers to fill power contracts.

Northeast of the Grand Canyon, the master stream of the plateaus is still the Colorado River, but the San Juan, Green, Gunnison, Dolores, San Rafael, Dirty Devil, Escalante and many other rivers have cut the country into a series of tabular blocks, separated from each other by what are for the most part non-traversable canyons. This erosion basin, of approximately one million acres, remains today as "fantastic, extraordinary, antediluvian" as it was when John Wesley Powell led his expeditions down the Green and Colorado Rivers.

Most of the 330,000 acres contained in Canyonlands National Park, created in 1965, have been leased for grazing for many years, and will continue to be grazed for another quarter of a century. But there are still places where, because of rough terrain or lack of water, no cattle have set foot. On the dry, windswept plateaus, the piñon pines and junipers are widely scattered, the latter often no larger than shrubs in spite of their great age. Some junipers, at the Black Canyon of the Gunnison, are more than 700 years old.

It is here and at the confluence of the Gunnison and Colorado Rivers that the Rocky Mountains meet the canyonlands. Near Grand Junction, the rock displacements associated with mountain-building have cracked the crust of the earth to form a 10-mile-long fault. The conspicuous escarpment, hundreds of feet high, runs through Colorado National Monument, where corridor-like canyons have been carved through the ages by the weathering of soft sandstone layers. As erosion proceeds, undermining the cliffs, huge blocks break off, deepening and widening the corridors. Fractures widen, leaving isolated columns and monoliths like Independence Monument, the first hint of the red-rock country to be encountered just across the Utah border.

In Colorado National Monument, rainfall is only about eleven inches a year. Under this semi-desert climate all broad-leaved trees disappear, except for water-conserving kinds—the single-leaf ash and curly-leaf mahogany—whose leaves are almost as tough and narrow as pine needles. Slim, gray-green ribbons of shrubs wind along drying stream beds in the canyon bottoms. Tadpoles swim in the shallow pools of the rocky stream-bed, every life cycle a rush to beat the drought. Above and below ground, ants swarm, and the pits of the ant lion border their trails. Lizards become as numerous as in the desert. Occasionally one can be seen fighting another, in a miniature version of dinosaurs battling to the death—as on the forgotten plateau of the Lost World. The aggressive leopard lizard kills and eats the side-blotched lizard, the sagebrush swift and the whiptail.

If you were to look for a male whiptail lizard

Natural Camouflage
Nearly always within one or two jumps of water, the canyon tree frog inhabits rock pools and large solution holes that fill up with water after a rain. Protected by coloration as effectively as any species, the canyon tree frog matches the rocks so closely that the animal is almost impossible to find.

in the Monument, your search would be doomed
to failure. The population is entirely female, and
the young are produced from unfertilized eggs.
This form of reproduction, known as parthenogenesis, is fairly common in the plant world and
among various invertebrate animals. In the last
ten years it has been discovered among some
fishes, amphibians and reptiles, including teiid
lizards in the New World and lacertas, their relatives in the Old World. The six or eight all-female
populations of whiptail lizards in the western
United States are the first of their kind known to
be self-perpetuating, with exactly the same genetic inheritance, generation after generation.

With unisexuality comes a slight difference in
ecological requirements. In the mountains of New
Mexico, the little striped whiptail, unisexual, occupies the grassland, oak and juniper zones.

Below, in the sand and mesquite, lives the New
Mexico whiptail, a spotted and striped, bisexual
species. That they have not been separated for
very long is proven by the fact that an individual
of the all-female species sometimes meets and
mates with a male of the other species in the
transition zone between 4,600 and 5,300 feet.
(The offspring of such a pairing are bisexual,
bigger than the parents, and faintly spotted.) If
this phenomenon occurs wherever ranges overlap
—and several other hybrids have been found—
parthenogenesis may offer the advantage of stability and survival to already established gene
combinations, *plus* the possibility of change.

With altitude, lizards and snakes generally
disappear, because of their dislike of the cold.
The only reptile that is really common in the
ponderosa pine on the higher plateaus of south-

An Unusual Lizard
Ranging from the lower sagebrush flats into the spruce and fir at about 10,000 feet, the short-horned lizard endures cold better than most reptiles. It has another peculiarity for a lizard; the young of this animal are born live instead of being hatched from eggs.

western Utah is the short-horned lizard, whose ubiquitousness over the Great Basin, the plateaus and the plains east of the Rockies, in a variety of environments, contrasts sharply with the isolation of the little-understood whiptails.

Following the roads up the canyons to the tops of mountains or plateaus in southwestern Utah, you can pass through several vegetation zones in the space of a few miles. The most abundant animals in the piñon pine are probably the rodents —pack rats, piñon mice, rock squirrels and cliff chipmunks. All of them thrive on the piñon nuts. The most assiduous collector is the pack rat, whose nest may contain as much as ten bushels of material, including not only an ample supply of piñon nuts but the special non-utilitarian items —the sticks, stones and bones—which this animal delights in accumulating. The elevation of fossil pack rat middens from the Wisconsin inter-

glacial period, some 10,000 to 20,000 years ago, indicates that—in Texas and California at least— the piñon forest moved downward 1800 to 2400 feet during this period, when the precipitation must have been much greater than today, and consequently all zones of vegetation lower. Over the millennia, this rodent and the trees on which it feeds have both persisted, simply by moving up and down a mountain. The pack rat is one of the successful survivors of the Ice Ages.

Glacial melt waters, which filled the Colorado and its tributaries, also stripped the plateaus of many rock layers during the "great denudations" that have exposed huge chunks of geologic time

The Virgin River
At Zion, Utah, billions of tons of rock have been ground up and carried away by the now-placid Virgin River and its numerous tributaries in Zion National Park.

in various localities. North of the Grand Canyon, erosion has created a series of escarpments, receding in a stair-step pattern into the distance—the Chocolate, Vermilion, White, Gray and Pink Cliffs. Each escarpment is developed on a resistant geological formation, ranging in age from 60 million to 200 million years.

Just as the Grand Canyon tells the story of the earliest eras, Zion, Cedar Breaks and Bryce Canyon in Utah fill out the story of the middle and late eras, as told by dinosaur prints and petrified logs. Here, hills of fossilized sand are broken by the cracks of plateau uplift, and rivers continue to carve canyons from the sediments of the ages. More than any other feature, the differential weathering of sedimentary layers—limestone, sandstone and shale—distinguishes these Utah canyons from others, some just as deep, in Idaho, Texas and California.

The Capitan Reef formation, which fronts on the deserts of New Mexico and Texas, was once a living reef in the Permian period, built by the calcareous algae and sponges of inland seas. Today, El Capitan Peak, the bold southern headland, is a monument to the extinction of these ancient sea creatures. The canyon walls at Cedar Breaks and Bryce Canyon consist of the silt and limy ooze that settled in the shallow lagoons of southern Utah much later, during Eocene times. The deposition of these strata (the Wasatch formation) is supposed to have ended about 13 million years ago; since then, the plateaus have risen from sea level to heights of more than 10,000 feet, and the steeper sides have eroded away into amphitheaters filled with pink limestone buttes and pinnacles. Dry as the arena seems, the architect was—and still is—running water. And the palette that gives brilliance to the spires is the same set of impurities in the rock that colors the Badlands of South Dakota and the Painted Desert of northern Arizona, regions also eroded by water. Here, slow-melting snow, which covers the tops of the cliffs to a depth of about three feet for several months a year, seeps into the cracks and dissolves the natural cement that holds the rocks together. Freezing and thawing opens out windows and arches from mere slits. Solution holes grow into caves. New sculptures are continually being created from the canyon walls as the figures in the basins are destroyed.

Hard, compact sandstones are acted upon by running water in much the same way. In Arches National Monument, where a 300-foot layer of the dull-red Entrada sandstone is exposed, some of the gigantic arches have been left isolated by the disappearance of all the surrounding rock.

Soft, porous sandstones absorb water and act as reservoirs, if underlain by harder rocks. Mesa Verde's Cliff House is a sandstone formation of this kind, in which the water-laden rock has peeled back to form natural caves and seeps—at once a home and a water supply for the ancient Indian Basket Makers who colonized the Four Corners area of Colorado, Utah, New Mexico and Arizona in the period 1 to 450 A.D. An entire Pueblo culture emerged on the mesa tops. Then, inexplicably, these people retreated from the mesas about 1200, and vanished a hundred years later. The multi-storied castles, nestled beneath overhanging cliffs, may have been deserted because of years of drought, and perhaps also as a consequence of long harassment by the Apaches and other nomadic raiders.

Human occupation of this country has always been as precarious as the ancient toe-hold trails over the cliffs, still visible next to Park Service ladders. Drought and impassable canyons make the high plateaus as wild as the mountain tops. Both are essentially roadless places, the last frontiers of wilderness, and as technology opens them up to mining, grazing, or recreational use, their fragile ecosystems may crumble.

One of the newest developments, and the one most difficult to assess, is the growth of a new kind of American nomadism, quite different from the nomadism in the arid lands of North Africa and Southwest Asia. In this country, a burst of road construction, plus mounting income and leisure time, have brought the opening up of large sectors of land previously untouched by men. In the 1950's, the Four Corners area was besieged by amateur uranium prospectors, and now a number of uranium "ghost towns" here are leading attractions on jeep tours. The new paved highways carry fleets of summer travelers, equipped with pop-up tents, trailers and other paraphernalia, all headed with single-minded determination for places far from the madding crowd.

In recent years we have discovered that the

fragility of nature is variable. Highly productive communities, such as the forests of the Pacific West, restore themselves rapidly after having been cut over. Many of the original plants come back, including a scattering of the rarer species. But in the less productive communities of the deserts, the mountains and the high plateaus, there are many plants with low growth rates and animals with low reproduction rates. These need protection, and a degree of protection that parks and refuges in themselves cannot provide.

Dams that alter the Grand Canyon can stifle a rhythm of life based on the flow of the Colorado over centuries of time, and flood or dry up the wading grounds of herons and the caves of desert bighorns. Water pollution has effects many miles downstream from industrial plants and mines, in what appears to be uninhabited country. Smog and fallout from nuclear explosions settles without regard to either the time or place of its origin. There are no longer boundaries between the natural and the man-made world.

Four Lanes to Everywhere
A new four-lane highway leads to a scenic viewing point below the sheer face of El Capitan Peak, rising 1,500 feet above a talus slope that contains marine fossils. Until recently the desolate and thinly settled region around the old reef in west Texas had few roads that were passable with an ordinary car. But massive federal and state highway programs are providing access to this and other previously remote places throughout the American West.

Acknowledgments

Without the understanding and special assistance of many people throughout the United States, this book would not have been possible. The authors are particularly indebted to Manuel Siwek, President of Grosset and Dunlap, Publishers, for providing funds and sufficient time to carry out the project. Financial aid for field research was supplied by the National Audubon Society and the Rockefeller Brothers Fund, through the auspices of Roland C. Clement, Vice President of the National Audubon Society. Special thanks are due to him and also to Gene Setzer, Executive Secretary of the Rockefeller Brothers Fund. Barbara and Roger Tory Peterson have given us notable help and encouragement throughout this long undertaking. Hobart M. Van Deusen, Head of the Archbold Expeditions of the American Museum of Natural History, contributed by assisting in obtaining a grant from the Explorers Club. Frank E. Egler of Aton Forest, Norfolk, Connecticut, read the manuscript in its entirety.

We are indebted to the following individuals attached to various divisions of the Department of the Interior. *In the Bureau of Sports Fisheries and Wildlife, Washington, D.C.:* Walter W. Dykstra, Division of Wildlife Research; Harry S. Goodwin, Chief of the Department of Endangered Species; Francis C. Gillett, Chief, and Phillip Dumont, of the Refuge Division. *In the National Park Service, Washington, D.C.:* Howard W. Baker, Associate Director (Retired), William C. Everhart, Chief of Interpretation, and William L. Perry, Conservation Education Specialist. *In the field:* Jack F. Welch, Associate Director, Jim Kennelly and Sam Linhart, Denver Wildlife Research Center; Wendell E. Dodge, Director, Forest Research Center, Olympia, Washington; Eugene H. Dustman, Director, Arnold L. Nelson, Mary and Bill Stickel, Fran Uhler and the entire remaining staff of the Patuxent Wildlife Research Center, Laurel, Maryland; Don Fortenberry, Rapid City, South Dakota (for special research on the black-footed ferret). Among the refuge managers and biologists who were especially helpful on our travels were Baine H. Cater, William L. French, H. C. Garrett, Milton K. Haderlie, Claude F. Lard, Eldon L. McLawry, Owen Vivion and Larry H. Worden. Among the naturalists of the National Park Service who spent a great deal of time with us, giving us the benefit of their personal experience in the field, were Tom Francis, Charles Parkinson, Jim Richardson, Paul Risk, Bill Taylor, Keith Trexler and Skip Wells.

Snow survey information was supplied by Morlan W. Nelson, Snow Survey Supervisor, Soil Conservation Service (USDA), Boise, Idaho.

For literature, transportation, guiding and other courtesies, we owe appreciation to the following state personnel: in California, Henry H. Hoover of the California Fish and Game Commission and Leonard Penhale of the Morro Bay Museum of Natural History; in Louisiana, Robert A. Lafleur of the Lousiana Stream Control Commission and Richard K Yancey of the Louisiana Wildlife and Fisheries Commission; in South Dakota, Bob Henderson of the South Dakota Fish and Game Commission.

For the opportunity to photograph and study desert wildlife and plants at the Arizona-Sonora Desert Museum in Tucson, we should like to thank William H. Woodin, Director, Lewis Wayne Walker, Associate Director, Rick F. Dyson, Merrit S. Keasey III, Mervin W. Larson, Paul W. Shaw and the rest of the staff.

Gratitude is acknowledged for the cooperation of the following individuals and organizations: The American Museum of Natural History Library, New York; John Borneman, National Audubon Condor Warden, Ventura, California; Peter W. Churchward, Dune Lakes, Ltd., San Luis Obispo, California; Don G. Davis, Director, Cheyenne Mountain Zoo, Colorado Springs, Colorado; Fred A. Folger, Museum of Science, Miami, Florida; Con Hillman, Rapid City, South Dakota; Mrs. Ross Luding, Sperry Chalet, Glacier National Park, Montana; John M. May, Curator-Manager, the May Museum of the Tropics, Colorado Springs, Colorado; the Miami Seaquarium; William A. Niering, Director, Connecticut Arboretum, Connecticut College, New London; Phil Owens, Corkscrew Swamp Sanctuary, Immokalee, Florida; Eugene and Steve Percy, Fillmore, California; Santa Barbara Museum of Natural History (Frederick Pough, Director, and Waldo Abbott, Curator of Birds and Mammals); Walter R. Spofford, State University of New York, Syracuse; Tall Timbers Research Station, Tallahassee, Florida; W. Verne Woodbury, Reno, Nevada.

Bibliography

General Ecology

Allee, W. C. and Karl P. Schmidt, *Ecological Animal Geography.* John Wiley and Sons, N.Y., 1966.

Allee, W. C. and Orlando Park, Thomas Park and Karl P. Schmidt, *Principles of Animal Ecology.* W. B. Saunders, Philadelphia, 1949.

Bates, Marston, *The Forest and the Sea.* A Mentor Book, New American Library, N.Y., 1961.

Craighead, John J. and Frank C., Jr., *Hawks, Owls and Wildlife.* Stackpole Co., Harrisburg, Pa., 1956.

Dice, Lee R., *Natural Communities.* U. of Michigan Press, 1952.

Elton, Charles S., *The Ecology of Invasions by Animals and Plants.* John Wiley and Sons, N.Y., 1958.

Elton, Charles S., *The Ecology of Animals.* Scientific Paperbacks and Methuen and Co., London, 1966.

Farb, Peter, *Living Earth.* Harper and Bros., N.Y., 1959.

Geiger, Rudolf, *The Climate Near the Ground.* Scripta Technica, Inc., *Trans.,* Harvard U. Press, Cambridge, Mass., 1965.

Kendeigh, S. Charles, *Animal Ecology.* Prentice-Hall, N.J., 1961.

Lack, David, *The Natural Regulation of Animal Numbers.* Oxford U. Press, N.Y., 1954.

Odum, Eugene P. and Howard T., *Fundamentals of Ecology.* W. B. Saunders, Philadelphia, 1966.

Shelford, Victor E., *The Ecology of North America.* U. of Illinois Press, Urbana, 1963.

Historical and Regional Descriptions

Austin, Mary, *The Land of Little Rain.* Anchor Books, Doubleday, Garden City, N.Y., 1962.

Bakeless, John, *The Eyes of Discovery.* J. B. Lippincott, Philadelphia, 1950.

Donnelly, S. J., *Trans. and Ed., Wilderness Kingdom; The Journals and Paintings of Father Nicolas Point, 1840–1847.* Holt, Rinehart and Winston, N.Y., 1967.

Farb, Peter, *Face of North America.* Harper and Row, N.Y., 1963.

Hastings, James R. and Raymond M. Turner, *The Changing Mile.* U. of Arizona Press, Tucson, 1965.

Hay, John and Peter Farb, *The Atlantic Shore.* Harper and Row, N.Y., 1966.

Jackson, Donald, *Ed., Letters of the Lewis and Clark Expedition 1783–1854.* U. of Illinois Press, Urbana, 1962.

Jaeger, Edmund C., *Desert Wildlife,* Stanford U. Press, 1961.

Krutch, Joseph Wood, *The Desert Year.* The Viking Press, N.Y., 1951.

Krutch, Joseph Wood, *The Grand Canyon.* Anchor Books, Doubleday, Garden City, N.Y., 1962.

Leopold, A. Starker, *Wildlife of Mexico.* U. of California Press, Berkeley, 1959.

Muir, John, *The Mountains of California.* Anchor Books, Doubleday, Garden City, N.Y., 1961.

Rand, Christopher, *The Changing Landscape.* Oxford U. Press, N.Y., 1968.

Steinbeck, John and Edward F. Ricketts, *Sea of Cortez.* The Viking Press, N.Y., 1941.

Teale, Edwin Waye, *The American Seasons,* 4 Vol. Dodd Mead and Co., N.Y., 1951–1965.

White, Laurence B., Jr., *Life in the Shifting Dunes.* Museum of Science, Boston, 1960.

Geology and the Distribution of Plants and Animals

Clark, Thomas H., *The Geological Evolution of North America.* The Ronald Press, N.Y., 1960.

Darlington, Philip J., Jr., *Zoogeography.* John Wiley and Sons, N.Y., 1957.

Dunbar, Carl O., *Historical Geology.* John Wiley and Sons, N.Y., 1965.

Gleason, Henry A. and Arthur Cronquist, *The Natural Geography of Plants.* Columbia U. Press, N.Y., 1964.

Mather, Kirtley F., *The Earth Beneath Us.* Random House, N.Y., 1964.

Scott, William B., *A History of Land Mammals in the Western Hemisphere.* Hafner, N.Y., 1962.

Sharp, Robert P., *Glaciers.* Condon Lectures, Oregon State System of Higher Education, Eugene, Oregon, 1960.

Thornbury, William D., *Regional Geomorphology of the United States.* John Wiley and Sons, N.Y., 1965.

Marine Biology

Carson, Rachel L., *The Sea Around Us.* Oxford U. Press, N.Y., 1961.

Carson, Rachel L., *The Edge of the Sea.* Houghton Mifflin Co., Boston, 1955.

Hardy, Alister, *The Open Sea.* Houghton Mifflin Co., Boston, 1956.

MacGinnitie, G. E. and Nettie, *Natural History of Marine Animals.* McGraw-Hill, N.Y., 1968.

Polikarpov, G. G., *Radioecology of Aquatic Organisms.* Scripta Technica, Inc. *Trans.,* Rheinhold Book Division, N.Y., 1966.

Ricketts, Edward F. and Jack Calvin (Revised by Joel W. Hedgepeth), *Between Pacific Tides,* Stanford U. Press, 1965.

U.S. Department of the Interior, *Fishes of the Gulf of Maine.* Government Printing Office, Washington, D.C., 1953.

Wimpenny, R. S., *The Plankton of the Sea.* Elsevier, N.Y., 1966.

Plants

Jaeger, Edmund C., *Desert Wildflowers.* Stanford U. Press, 1964.

Lemmon, Robert S. and Charles C. Johnson, *Wildflowers of North America.* Hanover House, Garden City, N.Y., 1961.

Martin, Alexander C. and Herbert S. Zim and Arnold L. Nelson, *American Wildlife and Plants.* Dover Publications, N.Y., 1951.

Sargeant, Charles S., *Manual of Trees of North America,* 2 Vol., Dover Publications, N.Y., 1961.

Vasil'yev, I. M., *Wintering of Plants.* Royer and Roger, Inc., *Trans.,* AIBS, Washington, D.C., 1961.

Animal Behavior and Adaptations

Bourlière, Francois, *The Natural History of Mammals.* Alfred A. Knopf, N.Y., 1954.

Busnel, René-Guy, *Ed., Acoustic Behavior of Animals.* ICBA-Elsevier, N.Y., 1963.

Carthy, J. D., *Animal Navigation.* Charles Scribner's Sons, N.Y., 1956.

Cott, Hugh B., *Adaptive Coloration in Animals.* Methuen and Co., London, 1940.

Hamilton, W. J., Jr., *American Mammals.* McGraw-Hill, N.Y., 1939.

Lorenz, Konrad, *On Aggression.* Harcourt, Brace and World, N.Y., 1963.

Milne, Lorus and Margery,

Patterns of Survival. Prentice-Hall, Englewood Cliffs, N.J., 1967.

Prosser, C. Ladd and Frank R. Brown, *Comparative Animal Physiology.* W. B. Saunders, Philadelphia, 1961.

U.S. Army and Navy Air Force, *The Physiology of Induced Hypothermia.* Proc. Symposium Oct. 28–29, 1955, NAS-NRC Pub. 451, Washington, D.C., 1956.

Withrow, R. B., *Ed., Photoperiodism and Related Phenomena in Plants and Animals.* AAAS Pub. No. 55, Washington, D.C., 1959.

Birds

Allen, Robert Porter, *The Roseate Spoonbill.* Dover, N.Y., 1966.

Austin, Oliver L., Jr. and Arthur Singer, *Birds of the World.* Golden Press, N.Y., 1961.

Banko, Winston E., *The Trumpeter Swan.* N. Amer. Fauna No. 63, Government Printing Office, Washington, D.C., 1963.

Bent, A. C., *Life Histories of North American Birds,* 20 Vol. Unabridged reprints of the U.S. National Museum Publication, Dover Publications, 1963–1965.

Brandt, Herbert, *Arizona and Its Bird Life.* The Bird Research Foundation, Cleveland, Ohio, 1951.

Brown, Leslie and Dean Amadon, *Eagles, Hawks and Falcons of the World,* 2 Vol. McGraw-Hill, N.Y., 1968.

Fisher, James and R. T. Peterson, *The World of Birds.* Doubleday, Garden City, N.Y., 1964.

Greenewalt, Crawford H., *Hummingbirds.* Doubleday, Garden City, N.Y., 1960.

Grossman, Mary L. and John Hamlet, *Birds of Prey of the World* (Photographed by Shelly Grossman). Bonanza Books, N.Y., 1968.

Hickey, Joseph J., *A Guide to Bird Watching.* Anchor Books, Doubleday, Garden City, N.Y., 1963.

Lanyon, Wesley E., *Biology of Birds.* Doubleday, Garden City, N.Y., 1963.

Lincoln, Frederick C., *Migration of Birds.* Doubleday, Garden City, N.Y., 1952.

Marshall, Joe T., Jr., *Birds of Pine-oak Woodland in Southern Arizona and Adjacent Mexico.* Cooper Ornithological Society, Berkeley, California, 1957.

Matthiessen, Peter and Ralph S. Palmer (Gardener D. Stout, *Ed.* and Robert Verity Clem, *illustr.*), *The Shorebirds of North America.* The Viking Press, N.Y., 1967.

McNulty, Faith, *The Whooping Crane.* E. P. Dutton, N.Y., 1966.

Pettingill, Olin S., Jr., *A Guide to Bird Finding,* 2 Vol., Oxford U. Press, N.Y., 1951–1953.

U.S. Department of the Interior (Joseph P. Linduska and Arnold L. Nelson, *Eds.*), *Waterfowl Tomorrow.* Government Printing Office, Washington, D.C., 1964.

Welty, Joel C., *The Life of Birds.* Alfred A. Knopf, N.Y., 1963.

Insects and Spiders

Evans, Howard Ensign, *Life on a Little-Known Planet,* E. P. Dutton, N.Y., 1968.

Frost, S. W., *Insect Life and Insect Natural History,* Dover Publications, N.Y., 1959.

Gertsch, Willis J., *American Spiders,* D. Van Nostrand Co., Princeton, N.J., 1949.

Needham, James G. and M. J. Westfall, Jr., *Dragonflies of North America.* U. of California Press, Berkeley, 1955.

Mammals

Dobie, J. Frank, *The Voice of the Coyote.* U. of Nebraska Press, 1961.

Einarsen, Arthur S., *The Pronghorn Antelope,* Wildlife Management Institute, Washington, D.C., 1948.

Ingles, Lloyd G., *Mammals of the Pacific States.* Stanford U. Press, 1965.

Jackson, Hartley H. T., *Mammals of Wisconsin.* U. of Wisconsin Press, Madison, 1961.

Linsdale, Jean M. and P. Quentin, *A Herd of Mule Deer.* Tomich, 1953.

Roe, F. G., *The American Buffalo: A Critical Study of the Species in Its Wild State.* U. of Toronto Press, Toronto, Canada, 1951.

Russo, John P., *The Desert Bighorn Sheep in Arizona.* Wildlife Bull. No. 1, State of Arizona Game and Fish Department, Phoenix, 1965.

Russo, John P., *The Kaibab North Deer Herd.* Wildlife Bull. No. 7, State of Arizona Game and Fish Department, Phoenix, 1964.

Walker, Ernest P. and Associates, *Mammals of the World,* 3 Vol., The Johns Hopkins Press, Baltimore, 1964.

Young, Stanley P. and Edward A. Goldman, *The Puma.* Dover Publications, N.Y., 1964.

Young, Stanley P. and Edward A. Goldman, *The Wolves of North America,* 2 Vol., Dover Publications, N.Y., 1964.

Young, Stanley P. and Hartley H. T. Jackson, *The Clever Coyote.* Stackpole Co., Harrisburg, Pa., 1951.

Reptiles and Amphibians

Carr, Archie, *Handbook of Turtles.* Comstock and Cornell U. Press, Ithaca, N.Y., 1952.

Carr, Archie, *So Excellent A Fishe.* The Natural History Press, Doubleday, Garden City, N.Y., 1967.

Oliver, James A., *The Natural History of North American Amphibians and Reptiles.* D. Van Nostrand Co., Princeton, N.J., 1955.

Pope, Clifford H., *Turtles of the United States and Canada.* Alfred A. Knopf, N.Y., 1949.

Index